Safety and Reliability of Programmable Electronic Systems

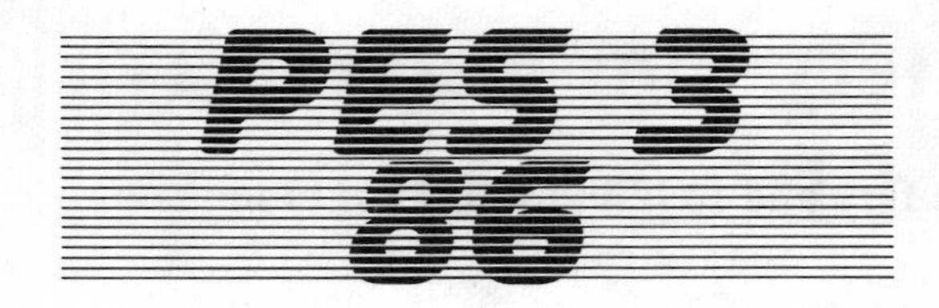

Proceedings of the Programmable Electronic Systems Safety Symposium held at the Beau Sejour Centre, Guernsey, Channel Islands, UK, 28–30 May 1986

Organised by
The National Centre of Systems Reliability, UK

Sponsored by
The Commission of the European Communities, Belgium
The Health and Safety Executive, UK
Arbejdstilsynet, Denmark
Berufsgenossenschaftliches Institut für Arbeitssicherheit, FRG
ElektronikCentralen, Denmark
Fraunhöfer Institut für Produktionstechnik und Automatisierung, FRG
The National Centre of Systems Reliability, UK
Institut National de Recherche et de Sécurité, France

International Programme Committee
B. K. Daniels (*Chairman*), UK
O. Andersen, Denmark
R. Bell, UK
H. Fangmeyer, Italy
J. Germer, FRG
K. H. Meffert, FRG
P. Nicolaisen, FRG
L. Steffensen, Denmark
B. Stevens, UK
J. P. Vautrin, France
A. M. Wray, UK

National Organising Committee
B. K. Daniels (*Chairman*)
R. Campbell
J. Peat
B. Stevens
D. Wolstenholme

Session Chairmen
H. Fangmeyer
B. K. Daniels
K. Meffert
A. de Kermoysan
O. Andersen
A. M. Wray
P. Nicolaisen

Safety and Reliability of Programmable Electronic Systems

Edited by

B. K. DANIELS

National Centre of Systems Reliability, Culcheth, Warrington, UK

ELSEVIER APPLIED SCIENCE PUBLISHERS
LONDON and NEW YORK

ELSEVIER APPLIED SCIENCE PUBLISHERS LTD
Crown House, Linton Road, Barking, Essex IG11 8JU, England

Sole Distributor in the USA and Canada
ELSEVIER SCIENCE PUBLISHING CO., INC.
52 Vanderbilt Avenue, New York, NY 10017, USA

WITH 22 TABLES AND 80 ILLUSTRATIONS

British Library Cataloguing in Publication Data

Safety and reliability of programmable
electronic systems.
1. Robots, Industrial—Safety measures
2. Robots, Industrial—Design and
construction
I. Daniels, B. K.
629.8′92 TS191.8

ISBN 1-85166-017-8

Library of Congress CIP data applied for

The selection and presentation of material and the opinions expressed in this publication are the sole responsibility of the authors concerned.

Printed in Great Britain at the University Press, Cambridge

PREFACE

The use of programmable electronic systems (PES) in industry has grown considerably with the availability of microcomputers. These systems offer many benefits to the designer and user in providing more comprehensive control of industrial processes, enviroments, machine tools and in robot installations. As confidence grows with the application of PES, users and manufacturers are considering incorporating safety functions within the requirements and functions of the PES.

This book represents the proceedings of the Programmable Electronic Systems Safety Symposium (PES-3) held in Guernsey, Channel Islands, May 28th - 30th 1986, which presented the guidance available to users, designers and safety assessors of programmable electronic systems. This guidance is applicable for many real and potential risk and safety situations in a wide variety of industries ranging from nuclear power plants and industrial robotics, to machine tools and chemical process controllers.

The original impetus to hold the Symposium came from a two year collaborative project partially funded by the Commission of the European Communities under the 1979-83 Informatics Initiative. The sponsors of the Symposium studied the assessment, architecture and performance of industrial programmable electronic systems, with particular reference to robotics. The group of papers in the first session give the first public report of the results of this project. The session was Chaired by H Fangmeyer from the Commission's Joint Research Centre at Ispra, Italy, who was the Commission's Project Manager throughout the collaboration.

Further sections of the book consist of a number of invited and contributed papers selected to reflect the many areas of concern in this area. Session II papers consider the safety and reliability of PES's in the operation of nuclear power plants while Session III covers experiences from other industries such as automotive systems and engineering components. Session IV deals with a number of methods for assessing the safety of PES's and in Session V the important topic of the reliability of the software controlling them is discussed. Following a session on the safety considerations of machine tools, the book concludes with a review of some important guidelines which have been drawn up by a number of National Governments.

Finally, I must thank all those who have assisted in the technical programming, the administration of the event and the preparation of these proceedings and particularly mention the members of the National Organising Committee and their staff.

B K Daniels

Contents

Session VI

EXPERIENCE WITH EMC, SIGNATURE ANALYSIS, FAULT SIMULATION AND SAFETY OF MACHINE TOOLS

Session VII

NATIONAL AND INDUSTRIAL GUIDELINES FOR PROGRAMMABLE ELECTRONIC SYSTEMS

List of Contributors

S N Ahmad 63
Head, Fuel Handling Controls, Nuclear Power Board, Department of Atomic Energy, Homi Bhahha Road, Colaba, Bombay-5, India.

O Andersen 27, 242
ElektronikCentralen, Venlighedsvej 4, DK-2979 Hørsholm, Denmark.

R Bell 1, 263
Health and Safety Executive, Magdalen House, Stanley Precinct, Bootle, Merseyside, L20 3QZ, UK

J Bøegh 242
ElektronikCentralen, Venlighedsvej 4, DK-2979 Hørsholm, Denmark.

A Bradley 152
AGR Systems Department, National Nuclear Corporation, Booths Hall, Chelford Road, Knutsford, WA16 8Q2, UK

G F Carpenter 164
Department of Electrical and Electronic Engineering, The University of Aston in Birmingham, The Sumpner Building, Costa Green, Birmingham, B4 7ET, UK

P V Castaldo 96
Senior Design Specialist, Nuclear Studies and Safety Department, Ontario Hydro, 700 University Avenue H9Bl, Toronto, Ontario, L4J 2W8, Canada

R Cipriani 105
IRT/SSC, Ansaldo Div Nira, Via dei Pescatori 35, 16123 Genova, Italy

B Clauzade 186
INRS, Electronique et Sécurité des Systèmes, 54501 Vandoeuvre, France

B K Daniels 41, 251
NCSR, UKAEA, Wigshaw Lane, Culcheth, Warrington, WA3 4NE, UK

D Dei-Svaldi 186
INRS, Electronique et Sécurité des Systèmes, 54051 Vandoeuvre, France

J Germer 36
Berufsgenossenschaftliches Institut für Arbeitssicherheit (BIA), Lindenstrasse 80, Postfach 2043, D-5205 St Augustin 2, West Germany

G Glöe 145
TÜV Norddeutschland, Grosse Bahnstrasse 31, D-2000 Hamburg 54, West Germany

R Greenberg 86
Licensing Division, Israel Atomic Energy Commission, P O Box 7061, Tel-Aviv 61070, Israel

W Grigulewitsch 176
Berufsgenossenschaftliches Institut für Arbeitssicherheit (BIA), Lindenstrasse 80, Postfach 2043, D-5205 St Augustin 2, West Germany

B Higel 186
INRS, Electronique et Sécurité des Systèmes, 54501 Vandoeuvre, France

D J Holding 164
Department of Electrical and Electronic Engineering, The University of Birmingham in Aston, The Sumpner Building, Costa Green, Birmingham, B4 7ET, UK

H Jansen 233
Head of Division, Electronics Railroad Technology and Vibration Control, TÜV Rheinland, P O Box 10 17 50, D-5000 Koln 1, West Germany

A de Kermoysan 122
Recherche et Technologies Nouvelles, Citroën Industrie, Direction des Equipments Industriels, 35 Rue Grange-Dame-Rose, ZI Velizy-Vitlacoublay, 92360 Meudon-la-Fôret, France

W G Kleppmann 117
Pruflabor für Elektronik, TÜV Stuttgart E.V., Postfach 2614, 7000 Stuttgart 1, West Germany

L A J Lawrence 74
Safety and Engineering Science Division, Atomic Energy Establishment, Winfrith, Dorchester, Dorset DT2 8DH, UK

F Manzo 105
IRT/SSC, Ansaldo Div Nira, Via Dei Pescatori 35, 16123 Genova, Italy

K Meffert 36, 176, 247
Berufsgenossenschaftliches Institut für Arbeitssicherheit (BIA), Lindenstrasse 80, Postfach 2043, D-5206 St Augustin 2, West Germany

K Natarajan 63
Fuel Handling Controls, Nuclear Power Board, Department of Atomic Energy, Homi Bhahha Road, Colaba, Bombay-5, India

P Nicolaisen 17
Fraunhofer-Institut für Produktionstechnik und Automatisierung (IPA), Nobelstrasse 12, D-7000 Stuttgart 80, West Germany

S Orme 74
National Nuclear Corporation, Booths Hall, Chelford Road, Knutsford, WA16 8Q2, UK

U N Pandey 63
Fuel Handling Controls, Nuclear Power Board, Department of Atomic Energy, Homi Bhahha Road, Colaba, Bombay-5, India

S P Petersen 242
ElektronikCentralen, Venlighedsvej 4, DK-2979 Hørsholm, Denmark

F Piazza 105
IRT/SSC, Ansaldo Div Nira, Via de Pescatori 35, 16123 Genova, Italy

D Pullen 74
Safety and Engineering Science Division, Atomic Energy Establishment, Winfrith, Dorchester, Dorset, DT2 8DH, UK

G Rabe 145
TÜV Norddeutschland, Grosse Bahnstrasse 31, D-2000 Hamburg 54, West Germany

V M Raina 96
Senior Design Specialist, Nuclear Studies and Safety Department, Ontario Hydro, 700 University Avenue H9 Bl, Toronto, Ontario L4J 2W8, Canada

G Reuss 176
Berufsgenossenschaftliches Institut für Arbeitssicherheit (BIA), Lindenstrasse 80, Postfach 2043, D-5205 St Augustin 2, West Germany

P Schreiber 133
Bundesanstalt für Arbeitsschutz, Vogelpothsweg 50-52, Postfach 17 02 02, D-4600 Dortmund 1, West Germany

A Schweitzer 199
INRS Service Electronique et Sécurité des Systèms, 54501 Vandoeuvre, France

I C Smith 74
Safety and Engineering Science Division, Atomic Energy Establishment, Winfrith, Dorchester, Dorset, DT2 8DH, UK

B Stevens 7
NCSR, UKAEA, Wigshaw Lane, Culcheth, Warrington, WA3 4NE, UK

M Sundquist 222
Machine Department, The National Board of Labour Protection in Finland, Uimalankatu 1, PL 536, Tampere 33101, Finland

J L Trassaert 210
Automobile Peugeot, Service INTX/ST, BP 50, 25207 Montbeliard, France

A M Tyrrell 164
Department of Electrical and Electronic Engineering, The University of Aston in Birmingham, The Sumpner Building, Costa Green, Birmingham, B4 7ET, UK

A M Wray 51
Health and Safety Executive, Research and Laboratory Services Division, Broad Lane, Sheffield, S3 7HQ, UK

ASSESSMENT ARCHITECTURE AND PERFORMANCE OF INDUSTRIAL PROGRAMMABLE ELECTRONIC SYSTEMS (PES) WITH PARTICULAR REFERENCE TO ROBOTIC SAFETY

R BELL

**Health and Safety Executive, Bootle,
Merseyside L20 3QZ, UK**

SYNOPSIS

The Commission of the European Communities (CEC) funded a joint project between seven organisations. The project ran for 2 years from September 1983. These organisations already had programmes of research in the fields of Programmable Electronic Systems (PES's) and robotic safety. The CEC funding provided the opportunity to collaborate in the exchange of information which would, hopefully, lead to harmonization of approaches to assessment of PES's in this field across Europe. The first part of this paper provides an overview of the project. Other papers given at this symposium will provide more detailed information about particular Objectives of the project. The second part of the paper looks at the future strategy for the use, development and review of the guidelines framework developed in the project.

INTRODUCTION

Computer based systems, generically referred to as Programmable Electronic Systems (PES's) are rapidly being deployed on a wide range of plant and machinery and this is creating a demand to use these systems for safety functions. Unfortunately, the rapid evolution of PES's has meant that in the safety context the store of experience and advice, which in non-PES systems has taken many years to accumulate, is not available to the extent that has traditionally been the case. Yet if this new technology is to be effectively exploited it is essential that those responsible for making decisions in this area have sufficient guidance on the safety aspects on which to base those decisions. A key element in the application of PES's for safety related applications is the assessment of the safety integrity that has been achieved in a particular design. The project was therefore essentially assessment orientated since PES assessment is a key issue in the effective exploitation of this technology in the safety field. The basic structure of the type of system coming within the scope of the project is shown in Figure 1.

CEC COLLABORATIVE PROJECT: BACKGROUND

The Commission of the European Communities (CEC) agreed to fund a collaborative project between seven organisations. The participating organisations where, in the United Kingdom, the Health and Safety Executive (HSE) who were the co-ordinating body, and the National Centre of Systems Reliability (NCSR); in Germany the Berufsgenossenschaftliches Institut fur Arbeitssicherheit (BIA) and the Fraunhofer Institut fur Produktionstechnik und Automatisierung (IPA); in Denmark the Elektronikcentralen (EC) and Arbejdstilsynet (AT); and in France the Institut National de Recherche et de Securite (INRS). The collaborative project identified a number of Objectives to be achieved during the two year project.

PROJECT OBJECTIVES

There were seven Objectives which, apart from Objective 7, were planned to be achieved within the two year period of the project. The Objectives were:-

Objective 1: Collection of information on PES's, specifically:-

(a) Collection of safety, reliability and availability data.

(b) Collection of accident and incident data.

(c) Examination of classification schemes in existence for the collection of data on the field operation of PES's.

(d) Analysis of the data obtained to provide a basis for the preparation of guidelines.

Objective 2: Creation of a databank

The data obtained in Objective 1 were to be classified and stored on NCSR's databank for analysis. If sufficient data were available, or could be collected within the time period of the project, the analysis would provide an important input into the formulation of the guidelines framework (Objective 5) and enable priorities to be determined. In the event that insufficient data could be collected, within the project timescale, the structure of the databank would enable future data collection activities in this area to be based on a common foundation.

Objective 3: Collection and assessment of current guidelines

Specifically:

(a) Collection of current guidelines for safe design, production, installation and operation of PES's. This included standards and guidance documents produced by international and national standards making bodies, safety regulatory bodies, professional organisations, trade organisations etc.

(b) Assessment of the guidelines, obtained in (a) above for their relevance in the context of robotic safety.

(c) Collection and assessment of current and proposed work that is directly relevant to the formulation of guidelines.

Objective 4: PES's in the context of robotic safety

To identify, with appropriate justification, those areas where further work was required. This involved the identification and examination of the specific areas, in the context of robotic safety, where the PES plays a role in the overall safety. Objective 4 therefore had to examine the results coming from Objectives 1, 2 and 3 and also those from Objective 5.

Objective 5: Formulation of guidelines

This Objective was to draw upon the work carried out in the previous Objectives and sought to:

(a) Formulate a technical guidelines framework for the immediate future to provide guidance on safety, reliability and availability assessments of PES's including both hardware and software. Emphasis was to be placed on quantitative techniques but it was recognized that in the immediate future qualitative techniques would have to be considered.

(b) Formulate future strategies for the use, development and review of the guidelines in (a) above.

Objective 6: Promotion of project results

Promotion of the guidelines and results obtained in the collaborative project to achieve, specifically in the context of the guidelines, acceptance by regulatory bodies, international and national standards organisations, designers, manufacturers and user organisations.

Objective 7: Seminar

A seminar was to be held at the termination of the project in order to publicize and aid the development of the results obtained. The overall management of the seminar was to be the responsibility of NCSR. (The PES-3-86 symposium is being held in order to meet the aims of Objective 7).

The Objectives were so structured that, in general, the work involved in the latter Objectives were based upon findings of the former. Objective 4 was exceptional in that it used results from Objectives 1, 2, 3 and 5. The main thrust of the project was concerned with PES's in general rather than concentrating solely on PES's associated with industrial robots. Objective 4 did, however, concentrate on the type of PES's used on industrial robots and considered the particular problems that were raised by the application of general guidelines (being formulated in Objective 5) to the type of safety problems that arise in the context of industrial robots.

Figure 2 provides an overview of the relationship between Objectives 1-5.

PROJECT OBJECTIVES AND THEIR RELEVANCE TO INDUSTRIAL ROBOTS

Most industrial robots are controlled by PES's. When assessing the adequacy of the safeguards provided on an industrial robot it may be necessary to consider the safety integrity of the PES based control system. In many cases it may be possible to adopt a safeguarding strategy which is independent of the control system eg by hardwired safety interlocks, physical barriers, mechanical stops to limit the arc of movement, by speed and torque limitation arranged by non-PES devices and by well defined and regulated systems of work. From a safety assessment viewpoint there is much to be gained by the adoption of safeguards that can be readily assessed and on which there is established guidance. However, there are situations where personal safety, minimization of damage to the industrial robot from adjacent structures and plant etc are dependent upon the continuing correct functioning of the robot control system. In some cases there may be economic penalties by adopting a safeguarding strategy independent of the control system. Whilst there are many advantages in using conventional safety devices, and this is to be strongly recommended in many cases, it would be unreasonable to prevent the use of PES technology for safety functions. A key issue is, as discussed previously, the problem of assessing the level of safety that has been achieved. An important aim of the project has been to develop a guidelines framework for PES's in general and to examine the implications of such guidelines when applied to a robot installation in particular.

DEVELOPMENT AND REVIEW OF THE GUIDELINES FRAMEWORK

There was a significant degree of interaction between the various Objectives (see Figure 2). In particular the development of the guidelines framework required input from Objectives 1-4. This meant that the framework that was developed had to be sufficiently robust to take into account wide ranging technical aspects as well as different national approaches to safety integrity issues. A framework so established is particularly suitable for use as a 'base document' for future national and international standards activities in this field.

In order to consider, in the context of the future use and development of the guidelines framework, the needs of designers, users and regulatory authorities, it is necessary that these groups play an active part in the future development of the framework itself and the application specific guidance developed within the framework. The mechanism available for such groups to play such a crucial role is already established if the development takes place under the auspices of both national and international standards organisations. In fact, the development of the guidelines framework, within the short and longer term, can only be done effectively through such organizations. Internationally, the International Electrotechnical Commission (IEC) is the most relevant organization within which to develop the work started in this project. The usual mechanism for bringing work of this nature to the attention of the IEC is to submit it via national standards organizations. If the guidelines framework was to be developed within the IEC, there would be an established mechanism for the review of documents developed in this way.

At the present time, there is no generically-based document, such as the guidelines framework, at an advanced stage in any International Standards forum. However, the Advisory Committee on Safety (ACOS) of the IEC has set up a Working Group with the following title and task:

TITLE: Functional Safety of PES's; Generic Aspects.

TASK: To describe the content of an IEC Publication "Functional Safety of PES's; Generic Aspects" and to prepare an elaboration for developing a guidelines framework for the application of programmable electronic systems (PES's) having safety functions. The envisaged publication should contain all recommendations which are not application - speciic but as a basis should:

1) Allow responsible Technical Committees to develop application specific recommendations.

2) Provide guidance on the application of PES's having safety functions in fields where no specific recommendations are made.

The setting up of this Working Group is an important development and the guidelines framework developed in this project should form a useful foundation on which further development can take place and help to achieve a common approach to PES assessment.

SUMMARY

For there to be sensible and effective exploitation of PES based technology, where there are safety implications in the event of PES failure, it is essential that the safety aspects are given proper attention. The CEC project examined several important elements of PES assessment - from data collection to the formulation of an assessment guidelines framework. This work should enable future decisions in this area to be based upon a sound foundation and hopefully provide a basis for a common approach to PES assessment within Europe.

If the work that has been done under the auspices of the CEC is to be effective in both the short and long term it is necessary to ensure that the ideas and views being developed, which are still at the formulative stage, are fed into the various national safety regulatory authorities and national and international standards making bodies. Only if this is achieved will the funding by the CEC have a lasting benefit to Europe.

FI/016/20-22/03-85/SHH

Figure 1: GENERALISED PES STRUCTURE

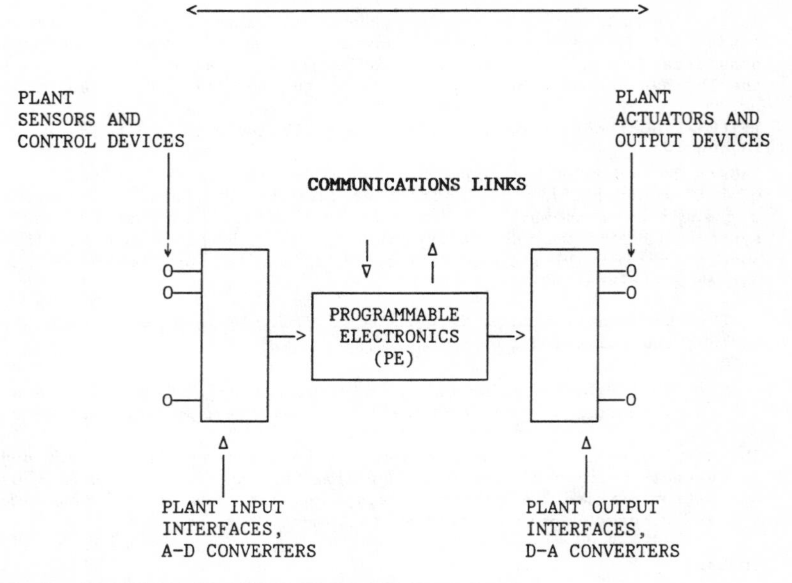

Figure 2: RELATIONSHIPS BETWEEN OBJECTIVES 1-5

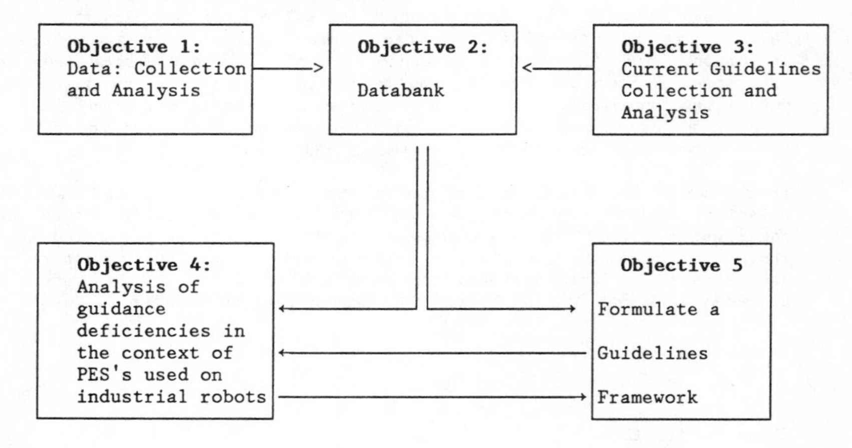

PRESENTATION OF OBJECTIVES 1 & 2 OF THE JOINT COLLABORATIVE PROJECT ON PROGRAMMABLE ELECTRONIC SYSTEMS COLLECTION & DATA BANKING OF INFORMATION

B. STEVENS

INTRODUCTION

This paper covers the work done during the project on Objective 1 - Data Collection - by Working Group 1, which had representatives from all the Project partners, and on Objective 2 - Data Banking - by The National Centre of Systems Reliability (NCSR) who were solely responsible for the data banking requirements of the project.

The technical annexe for the project required that Objective 1 undertook the following tasks:-

(1) Define specific areas of interest

(2) Review possible data sources

(3) Determine what data has to be collected and the method of collection to be used

(4) Define an "event" in the context of this project

(5) collect event data

(6) Examination of classification schemes in existence that may be suitable for use in this project. This is really an informatics problem as classification schemes are primarily used for the coded storage and retrieval of data in computer systems

(7) Analysis of the data collected by Working Group 1.

The technical annexe required that NCSR provided all data banking facilities as identified by Working Group 1. There is obviously a strong interaction between these two objectives, hence the combined coverage of these activities.

The seven activities set for Objective 1 will be covered separately with brief overview of the comprehensive data banking facility provided for the project.

OBJECTIVE 1

Definition of Areas of Interest

Before data collection could commence it was essential to define the areas of PES applications, including the projects declared bias towards industrial robots, that should be the targets for data collection.

The working group identified many areas where the current state of the art in PES was being exploited at a very fast pace and where there was a high likelihood of system interaction - for example in Flexible Manufacturing Systems (FMS). It was necessary, however, to be selective in the choice of target areas. This was done by considering the current fields of activity that the Partners were engaged in as a result the following three areas were selected:-

(1) Industrial Robots

(2) CNC and NC machine tools

(3) Automatic Warehouse Systems.

Review of Data Sources

Working within the confines of these defined areas Members submitted samples of data to enable The National Centre of Systems Reliability (NCSR) to undertake a data analysis. As NCSR were responsible for Objective 2 they required this to be done to enable work on the design and implementation of the data base to commence at an early date.

Sample data was supplied within the following categories:-

(1) Details of accidents that are submitted by industry to statutory bodies and insurance institutes.

(2) Information published in technical journals and conference proceedings.

(3) Results of direct mail questionnaires to PES manufactures and users.

(4) From access to PES users operational records.

(5) From access to PES manufacturers warranty and service records.

(6) Newspaper reports.

(7) Verbal or hearsay reports of events.

(8) Independent observation of system operations.

(9) Any other possible source.

Review of Existing Data Stocks

Sample data were received from all the Project members it ranged from current and historic press cutting to detailed reports into accidents involving equipment that had some part of its function controlled by a PES. The varied nature and absence of any population data devalued this information.

It was suggested that various data banks in Sweden could be a useful source of data. In particular those operated by the Swedish Labour Inspection Service at Vasteras and ASEA the robot manufacturers were identified as possible sources of data. Contact has been made with both and a visit was made by two members of WG1 to negotiate access to the data. The data in these banks is structured to provide statistics on lost time accidents at the work place and mainly provided information on the extent of injury and time lost rather than full details of the system involved and the real cause.

Data thought to be appropriate for the project was retrieved from the NCSR Reliability Data Bank.

The following general conclusions can be made on the quality of available data:-

(1) The sample data received from the partners in this collaborative project were very sparse in quantified parameters.

(2) It was mainly concerned with accidents and incidents.

(3) Only the existing NCSR data was directed towards reliability and availability and this was to some extent dated as there have been considerable changes in technology since these data were collected. The only exception to this was data coming from a current NCSR project.

(4) No partner submitted any data which would enable populations of similar systems to be estimated, this information would be required for statistical evaluation of any results.

Concurrent with the above activities NCSR conducted their own review of potential data sources viz:-

(1) Technical periodicals.

(2) On-line data bases.

(3) Machine Tool Industry Research Association (MTIRA - UK).

(4) Potential data collection sites in the UK and amongst Associate Members of The Systems Reliability Service (SRS).

(5) Manufacturers literature.

(6) Universities and Polytechnics (UK only).

Definition of Data to be Collected

A data analysis was carried out by NCSR using the sample data and predictions of the eventual uses to which the data base could be used to service the requirements of the other Objectives of the Project. This resulted in the following subject areas being identified:-

Events

This subject area will contain information relating to accidents and incidents in the previously defined areas of interest. For the purpose of this project the distinction between an accident and incident was defined as:-

"an accident in an event which may involve material damage to the system or product which results in injury to persons"

whereas

"an incident is only material damage to the system, product or a near miss but no personal injury".

It was the aim that sufficient information would be collected to enable quantified reliability and availability parameters to be derived. In fact during the life of the project only a few hundred events were collected and because these were derived from historic records there were large gaps in the data sets. This made the production of quantified parameters impossible.

Names and Addresses

This subject area contains names and addresses of people, companies and research institutes active in the defined fields. It was felt that this was required to enable direct contact with individuals who might have been able to contribute to the project.

Bibliography

This follows normal library style, the content was limited to books, reports and conference papers that are directly applicable to this project. Several searches of on-line systems have produced large hit lists, but subsequent detailed examination has shown a large proportion of non relevant publications. Input of data into this section was discontinued in March 1985 due to lack of response.

Standards and Codes of Practice

This area concentrates on recording those standards identified and reviewed by Working Group 3. The details of the assessments will be held on the computer giving the opportunity to do complex searching and sorting of these results.

Initial analysis of the assessments has enabled minor discrepancies to be corrected. Various matrices have been produced which serve to highlight those areas which are not covered by existing guidelines or specifications. Work is continuing in this area to enhance the facilities. The Appendix to the final report on Objective 3 was generated directly from this section of the Data Bank.

Text and Abstracts

It was intended that this data area would be a free text file system to contain such diverse published matter as newspaper reports, technical journal articles, verbal reports etc. Special attention would have to be paid to applying some form of grading to the contents to identify those entries which are substantiated and those that are apocryphal.

The nature of information received was very diverse and as such was extremely difficult to classify for the purpose of a computer based system.

Quantified Reliability Data

It was intended that summary reliability and availability data in tabulated format would be derived from some of the data source already mentioned in practice this was not possible as the data was very sparse and lacked homogenity.

Data Collection

The results of detailed investigations into all the possible data sources showed that the acquisition of data would need to be formalised to enable us to obtain the best quality and most appropriate information to fulfil the Objectives of this Project.

In the main it also appeared that data collection would have to be a primary function of the Partners, as no ready source of data had been identified.

Data collection forms were designed and covered the data subject areas 1-5 above, trial completion of the forms was undertaken and revisions made on the basis of the trial results.

The use of structured forms enabled uniform reporting from each organisation.

Classification Schemes

There are many industrial classification schemes in existence, their purpose is to enable a coherent selection, merging and manipulation of information within a data base. They should also be looked on as a means of transferring data between different data banks. If established or internationally accepted systems are used then the necessity of code translation by means of look-up tables is avoided.

The other major factor to consider in the adoption of any existing classification scheme for this project is that under Objective 2 of the collaboration the resultant data is to be banked by NCSR On their computer system. NCSR currently maintain very extensive classification schemes and also co-operate on an international basis in the development of new classification schemes.

The current practice in data banking is to only resort to coding of information where:-

(1) This preserves the homogenity of the data

(2) Security or confidentiality of the data is requried

otherwise information is stored in its literal form.

Where coding of informtion was deemed essential by either the NCSR Data Bank, for operational reasons, or the donor of the data, for security reasons, two existing schemes were used:-

(1) The NCSR Reliability Data Bank codes - these codes are not published and were used when it was necessary to maintain a high level of confientiality over the data.

(2) "Reference Classification concerning Component Reliability" published by CEC JRC ISPRA - No SA/+.05.01.83.02 on behalf of the European Reliability Data Banks Association (EuReDatA). This work reflects views, classification and standards being developed under international projects (OSO, CEE, IEC etc) and also the views of reliability and informatics experts alike.

It contains almost all the codes that were required for this project. If additional codes had been required then these would have been introduced to the Reference Classification via NCSR's membership of EuReDatA.

Analysis of the Data Collected

A detailed and subjective analysis of the data was not possible because of the lack of suitable population data and comprehensive details of the events for which we had knowledge. However, examination of the data that we had on a subjective basis suggested that the following conclusions could be made:-

(a) That there was no strong correlation between the occurrence of an 'event' and the control of the system by a programmable electronic system.

(b) In events where a worker was injured there was high probability that human error was the principal cause of the accident.

(c) In other areas where programmable electronic systems are used, in particular the process industries, the failure of these devices has led to near accident situations or very large economic losses. This is an area that we feel is worthy of further investigation.

OBJECTIVE 2 DATA BANKING

Overview of the Facilities Provided

The early results from the work done by Working Group 1 as detailed in 2.1 indicated that any data resulting from this Project would not be in a form suitable for direct entry and processing by NCSR's existing data bases. A decision was made to design and implement an entirely new and independent data base specifically for this project.

The new system is resident on a Fortune 32:16 super micro computer operating under 'unix'. The data base management system chosen is 'Informix' which is a relational type. The relational data base model is based on two dimension tables, all the records in a single table are of the same type. The relationship between records of tables are not held in lists of pointers but by holding common values in each related table. The rows of the tables are the records and the columns the attributes or fields. All records of one table have the same structure.

As all relationships are contained in the data itself and not in pointer structures modification to the initial data base design can be undertaken without affecting existing application programs. This was an important consideration as the exact nature and extent of the data base requirements were not known at the beginning of the project. During the life of the project the data base was subjected to several major modifications without any loss of existing data or the rewriting of application programs.

The use of Informix allowed us to design a system which the user accessed through a menu system to carry out all their main tasks eg:-

(a) adding new data

(b) modifying existing data

(c) retrieving data

(d) printing reports based on content of the bank.

The 'Perform' module of Informix was used to create the custom designed screen formats required to achieve this degree of sophistication for a sort term application. Figure 1 shows the interaction of the 'Informix' modules. The main screen of the system is reproduced in Figure 2.

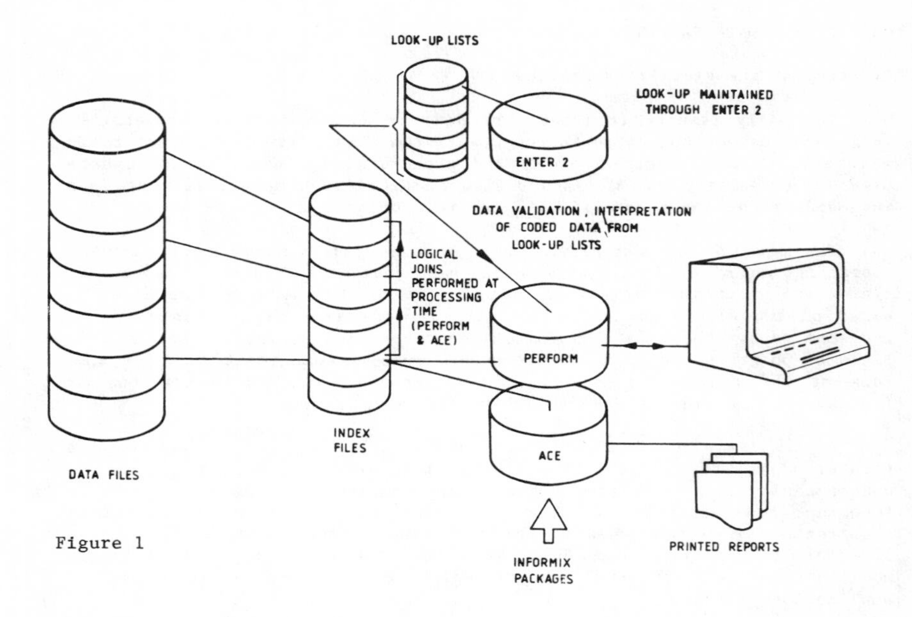

Figure 1

SCREEN 1 - MENU OF FACILITIES

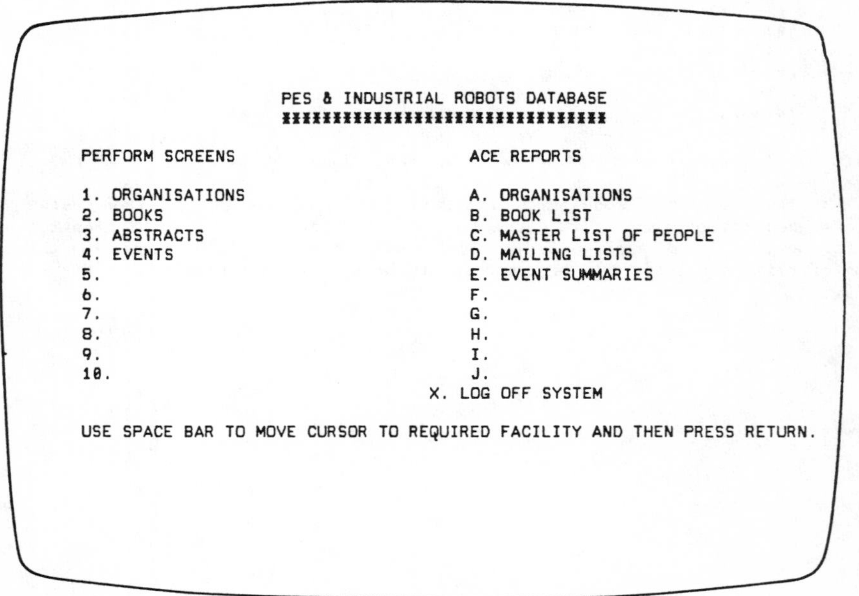

Figure 2

It is not appropriate to go into the full details of the design of the system in this paper, however, one feature we introduced may be of interest. As a large volume of data came from BIA and IPA in Germany we designed two sets of screens one in English and the other in German. The data was originally keyed in via the German screens and retrieved via the English screens using translation tables. The English version text was input at a later stage after translation. This was done because at the time it was thought that a very large volume of data would come from Germany and we would make the system available to the via the PTT. Facsimiles of the two of the screens are shown in Figures 3a-3b.

```
Query Next Previous Add Update Remove File Screen Current Master Detail
Output Bye                                   ** 1:accidents file **

                         ACCIDENT DETAILS

Type of event [ ]     Reference number [0    ]      Year: [     ]
Location:                  Location information:-

Source:- Code [  ] Info
Research codes: [  ][  ] Equipment a) Major code [  ]
b) Minor group - Document No [         ]    Description:-

Manufacturer: [ ]   Operator: [ ]    Activity code: [   ]
Equip response code: [   ] [   ]
Number injured: [  ]      Number dead: [  ]     Nature of injuries: [   ]
Causes    a) major [  ][  ]  b) minor 'technology' [  ][  ]  c) minor human [  ]
Component failures: [   ][   ]           Component failures description:

Environmental failures: [   ]     General remarks:-
```

Figure 3a

```
Query Next Frevious Add Update Remove File Screen Current Master Detail
Output Bye                                    ** 1:accidents file **

                              ACCIDENT DETAILS
Art Des Ereignisses:[ ]       Laufende Nummer: [0     ]
Ort Des Ereignisses:[                ]     Weitere Angaben Vorhanden:-
[                                                                    ]
Zeitpunkt Des Ereignisses: [     ]   Quelle: [  ]
Weitere Angaben Vorhanden: [                                         ]
Untersuchung Durch: [  ]  [  ]
Einrictung (Anlage/Gerat) a)Obergruppe: [  ]  b)Untergruppe: [          ]
[                                                                    ]
Herstelier Der Einrichtung: [ ]       Betreiber Der Einrichtung: [ ]
Tatigkeit Der Betroffenen person (en):[    ] Reaktion Der Einrichtung:[   ][   ]
Bei Unfallen a) [  ]      b) [  ]     c) [  ]
Storfall-Unfallursacne a)Obergruppe [   ][   ] b)Untergruppe "Technick" [  ][  ]
c) Untergruppe "Menschliches Verhalten" [  ]
Tecnologie Der Bauteile Bei Bauteile: [    ][    ]  Weitere Angaben Vorhanden:-
[                                              ]
Fehler Durch Umgebungs-einflusse: [    ]     Bemerkungen:-
[                                                                    ]
```

Figure 3b

CONCLUSIONS

The volume of relevant data collected during the project was disappointing. This was due to the time required to set up structured data collection campaigns. As initiatives Working Group 1 had made began to bring in results the project ended. It is evident that operators perception of industrial robots and other systems controlled by PES are a major cause of accidents particularly to maintenance workers. If the work started under this project was continued it could provide the information required to improve this situation.

ANALYSIS OF ACCIDENTS AND DISTURBANCES INVOLVING INDUSTRIAL ROBOTS

NICOLAISEN, PETER

Fraunhofer Institute for Production Technology and Automation (IPA)
Stuttgart
Germany

Industrial robots as a symbol of new technology with programmable electronics

There are no doubt more weighty safety problems in today's industrial society than those involved with the use of industrial robots. One only has to think of road traffic, nuclear power stations, environmental pollution or the problem of "hazardous substances". But this should not be allowed to detract from the question of safety while using industrial robots.

On the one hand, the industrial robot, in terms of numbers in use and different types of application, is at the very beginning of its development, i.e. the safety problems can only get bigger, both qualitatively and quantitatively. On the other hand, the value of the industrial robot as a symbol of technological progress should be taken into consideration (both in the positive as well as the negative sense). The over-enthusiasm, which has in recent years brought the industrial robot to the forefront of public awareness, could just as quickly turn into general condemnation, with the instrument previously known as a "job killer" becoming thought of as a "killer of men".

Does the industrial robot represent a safety problem?

Accident risks due to industrial robots result from the actual reason for their use, the moving of loads, i.e. work pieces or tools (grinding wheels, welding tongs, paint-sprayguns etc.),

- at high speed,
- in large movement areas,
- on partially intermeshing tracks,
- and in complex patterns.

In addition, it happens that sudden changes in the program are carried out, or parameters (position, speed, acceleration) may be changed as a result of external signals.

If it is hardly possible even in the normal situation therefore for an outsider to foresee the next movement, it is even more of a problem when, perhaps during a breakdown (e.g. in the path-measuring system or in the speed monitoring), completely unpredictable movements at an undefined speed occur within the (kinematically possible) movement area (Fig. 1).

Fig. 1 Transversing IR for the arc welding of large frames

It cannot be expected that these problems are only temporary in nature and that they will in some way solve themselves. Up to the present, the industrial robot has initially been used in relatively few fields and often untypically, in mass production conditions (scarcely used flexibility), so that the technical merits - which nevertheless at the same time represent the special hazard - frequently are not utilised. Yet even here serious changes can be seen which also have their effects on safety, e.g. because of the fact that:

o as a result of the increased use in small and medium series production, the extent of the programmed tasks will increase (more frequent contact between man and industrial robot),

o the ranges of tasks are becoming greater, so that the whole installation becomes more complex and therefore, in certain circumstances, less supervised and more prone to breakdowns, i.e.:

- more complex programs,
- several industrial robots working together,
- use of sensors (external data command),
- use of gripper-/tool-changing systems,
- travelling industrial robots,

o the energy potential present in the system grows or new energies are added:

- higher-performance industrial robots,

- the use of high-speed tools,
- industrial robots for cutting by laser, water jet or plasma.

Accidents and disturbances involving industrial robots

It is difficult to get hold of information concerning accidents and near-misses with industrial robots. About five years ago, some data on this subject were collected and evaluated at the Institute for Production Technology and Automation (IPA), Stuttgart. There were about 180 references to accidents and other problems, but they were of very variable quality.

An example of this is shown in Table 1, which uses Swedish data as a basis for a few typical descriptions of accidents and their consequences.

Although these Swedish data are not necessarily representative of the situation in other countries, the following statement could be applicable:

Accidents at work places where there are industrial robots occur less because of any individual, spectacular event, than, usually, because:

- various, frequently quite ordinary errors are made during design, at the planning stage of the installation, or in the organisation of the operational planning, which could in other cases not give rise to dangerous consequences, but lead to the accident in a work place with industrial robots,

- frequently far too careless handling of a machine which is in itself dangerous is permitted.

In most cases these involve two-person groups (Table 2), the programmer and the fitter and the machine operator and the person in charge of the plant during the clearing of a fault.

Table 2 Degree of risk to various groups of persons when IR's are used

Group at risk	% frequency
1. Programmer / fitter	57
2. Staff for clearing stoppages	26
3. Maintenance staff (repair/maintenance)	4
4. Operating personnel (normal operation)	13
Total	100%

Basis: 131 cases Source: IPA 1984

The industrial safety problem

It is a fact that even at the present time it is easier to describe the problem and to indicate errors and points of weakness, than to suggest

Table 1 Accidents at work places where there are industrial robots in a Swedish company

Triggering effect	Result	Injury
Break of the cable strain relief, causing cable damage	IR travels at high speed against end stop	Physical injury (arm)
Break of the cable strain relief, causing cable damage	IR remains standing under closing press tool	Material damage
Short circuit in hand programming device because of weld spatter	IR moves unexpectedly	Physical injury (arm and upper body)
Programmer presses wrong button (movement direction) when he changes position	IR travels in wrong direction	Physical injury (arm and upper body)
Programmer presses wrong button (movement direction)	IR travels in wrong direction	Material damage (IR)
Programmer selects incorrect creep speed	IR travels too fast	Physical injury (arm and upper body)
Machine operator puts hand into moving plant because work piece is in wrong position in the press	Collision with IR and press tool	Physical injury (finger severed)
Machine operator puts hand into moving plant because piece to be ground is wrongly positioned in gripper	IR presses hand against grinding wheel	Physical injury (back of hand abraded)
Machine operator frees jammed work piece in magazine	Gripper squeezes hand	Physical injury (hand)
After correctly switching off plant part of the system remains under pressure	Nozzle opens and sprays oil at high pressure under skin	Physical injury (finger amputated)
Machine operator puts hand into moving plant because work piece is wrongly positioned in the gripper	Gripper squeezes hand	Physical injury (broken finger)
Control error (subsequent reconstruction of event without result)	IR starts suddenly and collides with programmer	Physical injury (broken ribs)
Air line to gripper interrupted	Gripper loses piece	Physical injury (foot)
Spring in gripper broken	Gripper loses piece	Physical injury (foot)

Table la Accidents at work places where IR's are in use

Triggering effect	Result	Injury/Period off work
Maintenance worker made a a trial run at the industrial robot following change of valves. Unexpected movement of one shaft since air was present in the system	IT collided with maintenance worker	Physical injury (hip, leg) > 7 days
Maintenance worker is in movement area of a robot with installation running (Workpiece-handling: insertion into a degreasing plant)	Maintenance worker is struck by robot and trapped	Physical injury (bones) 12 days
Machine worker (lathe with HHS) is struck by handling device during setting up work	Hand is pressed against sharp working tool	Physical injury (finger) > 7 days
Machine operator attempted to release trapped workpiece in the tool, which was fed in by conveyance equipment, while plant (press) was running. As he released workpiece the conveyance equipment started up again	Hand was trapped	Physical injury (hand) 12 days
Machine operator attempted to lift a dropped part at a forge working place with 2 HHS at the same time as the second handling device tried to pick up the part	Handling device collided with worker	Physical injury (head, brain) > 7 days
Machine operator reaches into running plant (press with handling device) because part is trapped in the grab	IR starts up; worker attempts to avoid and injures himself at the workpiece-magazine	Physical injury (finger) > 7 days
During setting at IR-workplace the setter placed workpiece on the conveyor belt. The sensor was thereby activated thus starting up the IR	Setter was pressed against the machine by the IR and trapped. Colleague shut down the IR via the emergency stop	Physical injury (back, chest) > 7 days
Machine worker at the IR-workplace with protective fencing and monitored gates enters plant in accordance with regulations in order to eliminate interference	IR starts suddenly and squeezes hand	Physical injury (hand) 9 days

Table 1b Accidents at work places where IR's are in use

Triggering effect	Result	Injury/Period off work
Machine operator was to exchange device (?) at a plant consisting of two units while the first unit was at rest. He failed to notice that the other unit was still in operation	Hand was trapped between HHS and machine	Physical injury (hand) > 7 days
Machine operator (press) tried to remove wrongly-positioned workpiece while plant was running. Resumption following removal of stoppage	Hand was trapped	Physical injury (hand) 1-7 days
During trial run of an IR the inspector is to hold still/ counterstop the equipment during movement	Hand is trapped between moving shafts (shaft/counter weight)	Physical injury (finger) > 7 days
Machine operator wants to remove wrongly positioned workpiece. Thereby he inadvertently activates an external switch	IR starts up suddenly through external signal and squeezes hand	Physical injury (finger) 2 days
Maintenance workers wants to start up IR-plant with the help of a colleague following repairs	Because of a misunderstanding the IR starts up and traps a hand	Physical injury (finger) 12 days
Whilst a machine in an IR-installation is started up, the IR (programmed) also starts up	IR collides with programmer	Physical injury (arm) 10 days
Machine worker wanted to remove workpiece whilst plant was running because the magazine was full	IR placed hot workpiece on the hand	Physical injury (hand) > 7 days
Machine worker enters a running plant (welding) in order to remove parts	IR hit worker on the head	Physical injury (head) 17 days
IR-arm had got stuck. Machine operator overrode emergency stop and entered the plant. Energy still present (compressed air) put IR in motion	IR-arm pressed hand against mechanical limit stop	Physical injury (finger) > 7 days
Machine operator wanted to remove (wrongly positioned?) workpiece from the plant	Hand was trapped	Physical injury (finger) 10 days

remedies which can be generally applied.

The problem has been faced at a relatively early point in Germany, and efforts have been made to seek solutions and to formulate minimum requirements. In 1977, after some preparation by the Handling Systems Association ("Arbeitsgemeinschaft Handhabungssysteme" - "ARGE HHS"), an independent VDI Working Group was set up. The draft guidelines which it produced (VDI 2853) from the year 1979 required a short time later intensive revision, which can probably be completed at the end of 1985.

For reasons of space we are not able to go further into the contents of this draft here, and all that we would say is: safety at the work place where there are industrial robots does not result from any single measure, but is rather the product of three factors which are mutually dependent:

- a safe industrial robot,
- safe design of the whole installation
- safe operational procedure.

Even the safest robot cannot yet guarantee a safe work place, since it still has to move and for that reason alone presents a hazard. Safe design of the installation cannot compensate for short comings in the operational procedure, when, for example, safety devices are made ineffective or weak points and gaps are not pointed out or removed, as the weak points listed in Tables 3, 4 and 5 show.

Table 3 Weak points caused by design which affect safety when using IR's

DESIGN
* Simple errors in control lead to dangerous system states
* Unsafe gripper design (especial with power failure)
* Inadequate strength of cables/hoses; poor laying
* Component failure (mechanical) leads to dangerous system states (valves, position sensors)
* Inadequate protection against environmental influences (weld spatter, dust, swarf, temperature, electromagnetic radiation)
* Failure of basic safety devices (e.g. emergency shutdown switch)
* No protection against unintentional activation of operating elements (knocking against, leaning on, dropping)
* Poor ergonomic design (IR, hand-programming device, operating desk) increases the likelihood of incorrect operations

Table 4 Weak points caused by planning

LAYOUT

* Poor spatial arrangement
  (confusion; possibility of collisions)

* Poor organisation of work
  (particularly when clearing stoppages and programming)

* Unsafe or confused linkage
  (interfaces between individual machines)

* Inadequate safety devices
  - faulty emergency shutdown circuits
  - insufficient guards
    * gaps
    * too low
    * too close to hazard points

Table 5 Weak points in operational procedure

OPERATION

* Inadequate removal of weak points
  (no measures set up for removal; no feedback to design and layout)

* Permitting working procedures which are counter to safety (particularly during stoppage clearance)

* Inadequate training of personnel

VDI Guideline 2853 means that in the very near future a certain standard for safety in work places where there are industrial robots will be laid down.

There has also been much activity at an international level.

Since 1981 there has been a three-nation working party on "Safety with industrial robots", with representatives from France, Great Britain and

Germany, which also maintains contact with other European industrial countries, e.g. Sweden.

Furthermore, an ISO Working Party (TC 184) was set up a year ago, which, in a sub-group is specially concerned with safety matters with industrial robots. And, not least, there is the "CEC Collaborative Project on the assessment, architecture and performance of industrial programmable electronic systems with particular reference to robotic safety". Welcome and important though these activities may be, there is also a series of further, parallel, tasks which still remain to be solved:

o <u>To create awareness of the problem</u>

In order to achieve any success it is essential first to make all those involved (manufacturer, user, designer, planner...) at first aware of the problem, in order to reach a situation where safety problems are given the same priority as questions of technology or commercial viability, so that it becomes a matter of course to involve them in the considerations at a very early point.

An appeal is made here to the planning departments of the machine manufacturers, and particularly to the first user, to consult without being asked with regard to the absolutely essential safety precautions, and also to remain persistent even if, on the commercial side, items are being sought where savings can be made.

o <u>To provide back-up</u>

Any insight and motivation is not very helpful if it causes difficulties in transforming the safety requirements into practical measures. Check lists have to be produced for this, measures, catalogues and case studies have to be compiled which can also provide support to the user with little experience.

Preliminary work towards this has already been carried out at the IPA, which will lead to:

- recognising possible problems in good time
  (check list: Hazards when using IR's),

- laying out the work places in accordance with safe practice
  (Examples and Rules for the Design of IR Work Places),

- evaluating, selecting and integrating appropriate safety devices for the particular application
  (Catalogue: Safety Devices for IR Work Places)

- and taking measures to reduce stoppages which require manual intervention
  (Collection of examples: design or organisation measures).

o <u>To push ahead with new developments</u>

Finally, it is important to develop tailor-made solutions for special problems which will fill certain loopholes which are at the moment still present (e.g. safety measures for the programmer, observation of working procedures at full operational speed in the immediate vicinity of the IR) or which will deal with future problems which arise from further developments in robot technology (use of optical and tactile sensors; use of gripper- and tool-changing systems; machining with high-speed tools; the working together of several industrial robots at one location; mobile industrial robots; laser cutting, plasma cutting or water-jet cutting; use of small computers).

CONCLUDING REMARKS

The subject of accident prevention at robot workplaces is far too complex to be dealt with adequately in this paper.

I should like to make the following concluding remarks:

A solution to the problems will neither arrive on its own, nor emerge from the transformation of any single phenomenal idea. As a rule it consists rather of a more or less successful combination of different partial solutions which have frequently emerged in laborious small-scale work by the groups taking part in the project. Every effort should be made to make this cooperation possible, to improve it, and to make it easier.

REFERENCES

1. Nicolaisen, P., Occupational safety with the use of industrial robots. wt Zeitschrift für Industrielle Fertigung, 74, (6), 1984.

2. Accident statistics of the Swedish Safety Authorities. Arbetarskydd-sstyrelsen, ASS, Solna, Sweden.

3. Barrett, J., Bell, R., Duelen, G. and Nicolaisen, P. Problems of occupational safety in connection with the use of industrial robots. Paper presented at the meeting of the British-German-French Working Group on Aspects of safety of industrial robots, Paris 1981.

4. Nicolaisen, P., Novel safety device for industrial robots and machines. Robot news international. IFS Publications, March 1983.

5. Bonney, M.C., Yong, Y.F. (editors), Robot Safety, Springer Verlag and IFS Publciations, 1985.

6. Nicolaisen, P., Standardization of safety requirements for industrial robots, wt Zeitschrift für Industrielle Fertigung, 74, (6), 1984.

COLLECTION AND ASSESSMENT OF CURRENT STANDARDS AND GUIDELINES FOR PES - CEC COLLABORATIVE PROJECT, OBJECTIVE 3

OLE ANDERSEN

ELEKTRONIKCENTRALEN
Hørsholm, Denmark

INTRODUCTION

As part of the CEC Collaborative Project: "Assessment, Architecture, and Performance of Industrial Programmable Electronic Systems with Particular Reference to Robotics Safety", a survey and analysis of existing standards and guidelines were performed. Participants in this part of the work came from:

- o Elektronikcentralen, Denmark
- o Direktoratet for Arbejdstilsynet, Denmark
- o Berufsgenossenschaftliches Institut für Arbeitssicherheit, W-Germany
- o Health and Safety Executive, United Kingdom
- o Institut National de Recherche et de Sécurité, France
- o National Centre of Systems Reliability, United Kingdom

Below is given a short account of the collection and assessment work, together with the results of the analysis. A full description may be found in the final report [1].

SCOPE

The scope of Objective 3 is given in the Technical Annex of the collaborative project:

Objective 3: Collection and assessment of current guidelines - specifically:

3.1 Collection of current guidelines for safe design, production, installation, and operation of PESs. This would include the collection of guidelines that have been produced by international and national standards-making bodies, safety regulatory bodies, professional organizations, trade organizations, other bodies etc.

3.2 Assessment of the guidelines, obtained in 3.1 above, for their relevance in the context of robotic safety.

3.3 Collection and assessment of current and proposed work that is directly relevant to the formulation of guidelines. This would include work being done in the international and national standards-making organizations, work being done under the auspices of the CEC, expert committees, international and national research programmes, government based schemes etc.

In Objective 3.1, such material is collected which has a standing as official guidelines, recommendations for - or requirements to - the safe design, production, installation, and operation of PESs. The material described in Objective 3.3 further includes proposed guidelines of the above mentioned kind together with other material which the partners find relevant in the light of the work performed in Objectives 4 and 5. This includes e.g. material giving guidance on the safe use of robots, without concentrating on the PES part, together with material of a tutorial nature, and descriptions of institutions working in the field.

In defining which material to include in Objective 3.3, no fixed requirements have been used. As an example, some guidelines on protective devices (e.g. photoelectric safety systems) have been included, but a systematic collection has not been attempted. Similarly, other material not being of primary importance or unique to PESs has only been covered sporadically.

COLLECTION AND REVIEW OF MATERIAL

The priorities in selecting and assessing the material as indicated above have been worked out based on the main aim of the collaborative project. This aim is stated as Objective 5 of the Project, requiring guidance on the safe use of PESs to be developed. Thus, a knowledge of already existing material will be very valuable, whether this concerns guidelines for use of PESs in e.g. medical instrumentation, industrial applications, or aviation instrumentation. This is due to the fact that many of the qualities needed for the safe use of PES hardware and software are largely independent of its application. Much relevant advice can therefore be found in the requirements to PESs used in the air traffic control, medical devices etc.

It should be noted that the search for guidelines has not been exhaustive, since all possible aspects, countries, and institutions have not been covered. Based on the partners' experience, an approach was made to those institutions that were expected to have most experience in the field of guidelines for PESs. A further knowledge of such institutions which have been working with similar guidelines shows where to seek additional information relevant to the work on Objective 5. A contact was therefore established with these institutions.

When the material has been collected, an assessment is needed, since only an analysis can show whether the quality is such as to make the material interesting.

In order to analyse the material and cover Objective 3.2 as well as the assessment part of 3.3, a profile of each document was established using a 3-page assessment sheet. The form used does not include a statement concerning the official status of the document in question, since the assessments are to be used as a data base for the Project's Objectives 4 and 5 only. Some of the documents assessed have a legal standing in some countries, some have no legal standing at all, and the status of some is uncertain.

The present assessment of the guidelines is based on easily identifiable criteria. A detailed assessment of the robotic safety aspects will depend heavily on the work of Objective 4, and will therefore be integrated with this work.

In searching for (inter)national guidelines, the following sources were exploited:

. national and international standards
. official publications
. trade organizations
. testing institutions
. manufacturers
. CEC publications
. IEC/ISO/EWICS publications
. PES manufacturers'/importers' advice to customers
. trade union publications
. professional organizations IEE/IEEE
. national and international research programmes

including the major institutions and organizations in Denmark, United Kingdom, West Germany, France, Italy, United States, and Japan , together with the other Nordic countries.

Such organizations, for which the partners lacked information regarding their work, were contacted by letter. Some of the organizations answered by sending relevant material, some indicated that they had no material of interest, and some did not answer. Representatives from a few of the most relevant organizations have been interviewed. A list of the document titles is given in APPENDIX A of the final report of Objective 3 in [1].

ASSESSMENT OF MATERIAL

An assessment sheet has been developed in order to ensure a uniform evaluation of the material. The front page gives the title, scope, and a short review of the document in question. The rest of the form contains a checklist for establishing a profile of the document coverage and specification level.

When applying the assessment form, the aspects of the systems considered in the guidelines were established first. This is done by determining whether the document is concerned with PES hardware, PES software, or non-PES aspects of the system. Based on the aspects considered, an interpretation of the rest of the items on the checklist may be made. In this way, items concerned with testing shall be understood as testing in the context of those aspects treated in the document.

The points of the checklist are, on a subjective basis, assigned an L, M, or H, according to the following definition:

[L] : The document mentions the problem without a proposed specification on measurement or assessment, or it describes the problem without many details.

[M] : The document gives a brief specification on measuring or assessing with some degree of subjective valuation, or it describes the problem in some detail.

[H] : The document gives specifications for objective measurement or assessment with a minor degree of subjective valuation, or it gives a very detailed description.

This grading system was adopted in order to establish a basis for the document assessment and, since it allows a quick retrieval of relevant documents, e.g. documents concerned with PES-testing may be identified very fast. The assessment results are shown in APPENDIX B, and a survey of the grades assigned for each item may be found in APPENDIX D, both appendices belonging to the final report of Objective 3 in [1] .

When a structured approach to the document assessment is applied in the way outlined above, the task of evaluating the document is divided into a series of subtasks. Hereby, the reviewer is led through a disciplined consideration of all aspects of the document.

When applying the assessment sheets, it was found that a certain amount of interpretation was necessary in order to use the form and grade system. This is due to the documents covering a wide range of subjects, from very detailed descriptions of robotics hardware, to standards concerned with the development of planning documents for the software development process. A further experience shows that the influence of subjective judgement is less on the grade assignment than on the text review of a given document. The text review is included on the front page of the assessment form.

In order to validate the assessment method, a cross check was performed by having a few guidelines assessed by different partners. A certain variation of the details of the grade assignments was found. The differences were mainly due to a few obvious misprints. It was noted, however, that the assessment form allows mistakes to be made, since for each point the assessor must make a relatively quick decision regarding the grade. The decision must be based on his recollection of the typically 10 - 100 pages of the guideline. An example of the variation introduced in this way is found by comparing two assessments. In these two assessments of the same document, several small deviations are found, where the two assessors have assigned grades differing by one level, e.g. H instead of M, or L instead of nothing. This just reflects the way marginal assignments are judged differently by the individual assessors. Furthermore, in both assessments, one box is found to have been left blank, where the other has the grade H assigned. A careful assessment shows that both boxes should contain an H. This gives an idea of the uncertainty present when using the assessment forms.

In conclusion, the validation shows that the assessments do give the reader a satisfactory identification of the aspects covered by the documents, and the level of detail to which the treatment is taken.

An enhancement of the assessment integrity is possible by making a full cross assessment so that each sheet has to be agreed upon by two partners knowing the material in question. The very different background of the partners implies that this approach would increase the assessment work by more than a factor of two, since the extra assessor would have to read all the material, of which some was already known to the original assessor. For this reason, and due to the different document languages, such an improvement was not attempted.

The content of the assessment forms has been included in the data bank, generated in Objective 2 of the Project. Further analysis, i.e. searching for specific assignment patterns, could be performed using the data base.

SUMMARY OF ASSESSMENT RESULTS

When interpreting the results from the assessment of the material it should be noted that the search was aimed at PES relevant documents concerned with hardware and software as well as documents relevant for robotics. This means that mandatory requirements in non-PES documents having an impact on PESs may easily have been overlooked. An example of this is found, if a document requires a mechanical device to be applied in a specific situation, thereby excluding the possibility of using a PES for that application. Another example is a general requirement, e.g. in a national law, that specific safety levels must be fulfilled in devices used in certain applications. Such a requirement may not have been found in the present search.

The assessment described above fulfils the requirements of Objectives 3.2 and 3.3. A final conclusion regarding the relevance of each document in the context of robotics safety will result from the work of the Project's Objective 4. The work in Objective 4 is concerned with the determination of those parameters of importance to the safety of systems in general, and specifically regarding robots. Thus, the results of Objective 4 will identify such requirements which must be included in a document in order to reflect the relevance in the context of robotics safety.

The following results were obtained:

- A total of 104 guideline assessments and institutions assessments were performed. Eight of the documents were found to give quantification of the safety integrity criteria.
- No guidelines, applicable to the entire field of interest and at the same time acceptable for all countries and applications, were found.
- Very few guidelines were found which had a scope similar to that of Objective 5.1 of the Project, and with a content close to the intended content of the documents to be developed in Objective 5.1.
- It is unlikely that any existing guideline will become accepted throughout EEC in its current form.
- The documents with the broadest coverage of the field of interest were of European origin.
- Most guidelines are not specifically addressing robotics systems, but most of the guidelines have some relevance to robots.
- Most of the documents are of interest mainly to designers and manufacturers. Only very few are relevant to users.
- Existing guidelines tend to concentrate on few aspects of the problem (hardware, software, or robotic-specific). Most guidelines are very limited as to their scope (concerned with e.g. failure detection and

localization in process computer systems, electrosensitive safety systems, or software documentation.

- o Significant gaps in the coverage of the problems were identified in the collected material. As an example, no fully developed operational quantification method was found to treat the question of software quality. The gaps in the material, in the context of robotics safety, will be identified in the final report of Objective 4 of the Project.

- o The existing guidelines employ a wide range of approaches, covering:
  - mandatory requirements
  - checklists with optional measures
  - content of planning documents e.g. quality assurance plan (the "what to do", not the "how to do it")
  - recommended practices

  The variety reflects the different national, legal traditions together with the different aims of the institutions developing the material. This means that in some countries, guidelines with a legal status exist. These guidelines tend to give general rules on what to achieve but they do not indicate how to achieve it, nor do they give advice to designers and producers. In other countries, standards exist which the producers may choose to apply. The standards have no legal standing, but they allow the producers to argue that they have followed "good engineering practices" in an eventual court case concerning questions of legal liability.

- o Many organizations have programmes concerned with developing guidelines and standards, although their specific aims may be different. Some may be concerned with safety at work, and some with standardization per se. Thus, the resulting documents are very different.

- o When regarding the entire set of guidelines, inconsistencies exist at some points. In some cases, a given system will fulfil the requirements in a guideline which deals with particular safety features at a higher or more general level. The same system may still fail to meet the requirements stated at a more detailed level in another document. Even more important is the fact that some documents disagree on what are the safe features of a system.

- o Although in practice, guidelines allowing a wide range of design strategies can be developed, some of the analyzed guidelines contained very specific - and design constraining - requirements.

- o Different large standardization organizations apply different philosophies in the standardization. As an example, take 4 standards concerned with software documentation. They have different requirements, which will force the manufacturers to develop different sets of software documentation, even if the information content is largely the same. No fundamental differences can be seen as to their overall objective. The variations are on an arbitrary level, but for a manufacturer of equipment to be sold in different market areas, the problem is serious.

  ANSI/IEEE 829:
  Specifies a fixed format and table of contents for the documentation.

DIN 66230:
Specifies what the content of the documentation should be. The format may vary.

BS 5525:
Suggests - by giving checklists - a flexible content, format, and documentation level. The checklists are used in order to ensure that the omission of a specific point is made deliberately, and not as a result of a mistake.

FIPS-PUB-38:
Specifies a number of documents and the table of content for each. Suggests a content of each chapter.

It should be noted that the documentation described in the four documents does not cover exactly the same areas, but the principle is demonstrated.

DISCUSSION

The results presented above are based on collection, analysis and evaluation of existing documents of relevance to the field, covering both official guidelines, draft versions, and material of a more tutorial nature. As a result of the collection, a good coverage of the existing material was obtained, even though some material of value may have been overlooked. It is the impression that the only major deficiency in the coverage may have been introduced by focusing the material collection on the documents that were directly associated with PESs. Hereby, requirements of a general nature in non-PES documents - but with a significance for PESs - may have been overlooked. On the other hand, the limitations introduced on the search were necessary in order to carry out the work. Without the limitation to PES-relevant documents, the workload would have increased at least an order of magnitude.

The most important result of the document analysis is the fact that, at present, no guideline exists which applies to the entire field of interest as stated in the technical annex. Different documents were found which cover different aspects of the system in question. What are lacking, are documents covering the entire field and putting the other documents into the right perspective. By doing this, such a document would define a framework of the complete view of the system. The document developed in Objective 5 of the Project will be an attempt to develop such an upper level document.

At present, it is difficult for designers, manufacturers, and users to choose a consistent set of guidelines, since the existing documents must be put together in a complicated jig-saw. This situation is aggravated by the above mentioned arbitrary inconsistencies introduced by having several guidelines covering the same aspect of the systems for use in different countries, different applications etc. Furthermore, a close inspection has revealed a number of gaps in the coverage of the guidelines - certain aspects of the system are not treated by any guideline. The identification of these gaps, in the context of robotics safety, are dealt with in Objective 4 of the Project.

Since the number of documents and draft documents has been seen to increase rapidly during the last few years, it is important that top level documents are developed very soon. If the development is not controlled, a situation will occur in which diverging requirements are developed in different countries, by different institutions, and for different applications. It should be noted that the formation of an overall framework is adversely affected by the existence of guidelines that are (partly) overlapping, contradictory, and inconsistent. This problem will increase as more and more special-purpose guidelines are introduced.

Large benefits to the CEC-community can be obtained by enhancing the harmonization across national borders in this field. The major effects will be the fact that the safety of systems can be enhanced without introducing barriers to trade. Further benefits from a general improvement to quality and reliability of PESs can be expected to emerge following an initial concentration on safety. A spin-off from this development will be decreasing development costs since the systems are met with consistent set of requirements. Further decrease in development costs can be expected with the structuring of the development effort. This will be caused by the introduction of existing guidelines, since many of these require the application of "good engineering practice".

Of special importance is the introduction of harmonized guidelines for PESs in major hazard areas due to the possible adverse effects on the popular acceptance of the new technology in case of serious and spectacular accidents.

RECOMMENDATIONS

The discussion above leads to the following recommendations:

- A top level document should be developed in order to facilitate standardization across borders. The document should contain a framework for the overall view of the system and put the existing documents into their right perspective.

- The above mentioned document should be developed within an international organization since the safety of PESs covers a large field of expertise. This development should be started at the initiative of an international committee and be parallelled in each country.

- Use of such a universal framework and a disciplined approach to system development should be supported by the CEC, especially for application in high-risk areas. The acceptance of the framework may be achieved by introducing an economic incentive.

CONCLUSION

This project has developed a cross-reference index for guidelines and documents concerned with assessment, architecture, and performance of industrial programmable electronic systems. The index and the related analysis of the material have been stored in a computer data base, and can be used by designers/manufacturers to gain access to existing guidance. A survey of the material may be found in the appendices of the final report of Objective 3 in [1]. If this material is to be kept up to date on completion of the project, it will require support to acquire and analyse guidelines worldwide and to maintain the data base. In case the recommendation above is followed, and an upper level document is developed, the data base will probably need to be updated.

The material collection and analysis effort has mapped out the entire field of existing material. It is the impression that only a small fraction of important material may have been left out.

The most significant result is the apparent lack of a top level document, containing a definition of the framework for the overall view of the system. A requirement to such a document would be to put the existing documents into their right perspective.

[1] CEC Collaborative Project "Assessment, Architecture, and Performance of Industrial Programmable Electronic Systems with Particular Reference to Robotics Safety", Final Report, 1986.

THE INADEQUACIES OF RESEARCH INTO PROGRAMMABLE ELECTRONIC SYSTEMS IN INDUSTRIAL ROBOTS

J. GERMER, K. MEFFERT

Berufsgenossenschaftliches Institut fur Arbeitssicherheit,
St. Augustin, Federal Republic of Germany

INTRODUCTION

This paper arose as part of the project "Assessment, Architecture, and Performance of Industrial Programmable Electronic Systems with Particular Reference to Robotics Safety". It presents the results of sub-project 4 the objectives of which were defined as follows: sub-project 4 is concerned with revealing areas which will require further work in the future. Of particular interest are those areas in which programmable electronic systems play a part in overall safety. The work of sub-project 4 was to be based on those results achieved in the preceeding sub-projects.

The original objective of sub-project 4 thus restricted work to those areas in which programmable electronic systems played a safety role. As work progressed, however, it became clear that when assessing safety as a whole, programmable electronic systems cannot be analysed in isolation but rather, that all contributory system elements must be taken into consideration. This was why the original area of analysis of sub-project 4, restricted to programmable electronic systems, was extended to encompass all those sub-systems and components of industrial robots which influence safety.

In the effort to achieve the stated objective not only the results of the preceding sub-projects 1, 2 and 3 were used but in addition a list of a few safety measures employed when using industrial robots was compiled and a few known and supposed problem areas concerned with robotics safety were set down along pragmatic lines. All participants in the sub-project were able to express their expert opinions on the individual subject areas on the basis of these findings.

Finally, sub-project 4 provided a description of a robot installation. This was meant for a case study planned for sub-project 5 on the application of basic guidelines developed in the plant. Any shortcomings of the basic guidelines were to be explored in an effort towards improvement. Inadequacies which could not be eliminated in this way were to be revealed in sub-project 4 as areas requiring further work in the future.

RELEVANCE OF PES TO INDUSTRIAL ROBOT SAFETY

Before focusing attention on the question as to which research inadequacies are still apparent in the area of robotics safety, sub-project 4 was initially concerned with the relevance of programmable electronic systems to industrial robotics safety. This was because sub-project 4 was the only one out of a total of 7 sub-projects which was to deal explicitly and exclusively with industrial robots. For this reason the subject of the significance of programmable electronic systems in industrial robots is given brief consideration in the final report on sub-project 4. Following on from this, the structure of industrial robot control systems are described beginning with the very first control systems using bit slice processors through to the modern multi-processor system of today.

Two comments made in the above-mentioned chapter of the final report on the trend of development in this area are of particular importance for robotics safety. It is expected that industrial robots will be used to an increasing degree in the future for small and medium-sized production runs and that the field of activity of the industrial robot will be extended with the result that the entire robot system will become more complex and thus more difficult to monitor. Both developments will presumably lead to a greater interaction between man and machine.

It is this very increase in the interaction between man and robot that leads us to conclude that the use of industrial robots in the future will entail increased risks if suitable guidelines and rules cannot be worked out prior to this development.

The possible dangers involved in the future and present-day application of industrial robots will be shown in the next section. It must be said, however, that the use of industrial robots does not only produce potentially dangerous situations but that a considerable degree of safety at work can be achieved through the application of this new technology. Having said all this there is still a necessity to develop certain procedures for the improvement of safety at work - also with reference to industrial robots. This is of particular significance in cases where dangers arise through the use of industrial robots which no-one had even thought of a short time ago.

HAZARDS ASSOCIATED WITH ROBOTS

In most cases industrial robots are not used alone but as part of a production environment consisting of a multitude of different machines. Human dangers exist in this production environment which can be instigated by the industrial robot itself, by the machines around the robot, by the interaction between industrial robot and the other machinery, by the material which is being processed by the robot or by the interaction between man and the industrial robot.

The high speed of movement of the industrial robot or the new and greater complexity of the sequence runs which lead to the loss of simplicity of operation for the user and service staff alike are examples of potential dangers of the kind mentioned above. Often, machines which have been constructed for robot operation are not equipped with the traditional safety precautions. This signifies an increased risk for the person who must repair the machine if a fault occurs.

When powerful processing tools such as water torches malfunction they can be dangerous far beyond the boundaries of the industrial robot's working area. In cases such as this people are possibly not as aware of the danger as they are when in the direct vicinity of the industrial robot during maintenance, repair and adjustment work.

Both these and many more dangers provoked by industrial robots are mentioned in the final report of sub-project 4.

IDENTIFICATION OF AREAS REQUIRING FUTURE WORK

We now come to the actual subject matter of sub-project 4, that of the identification and analysis of areas which will require further work in the future. First of all, those acts, regulations, guidelines, directions etc., collected as part of sub-project 3 which are connected with programmable electronic systems were examined. It became apparent that quite considerable gaps exist in the official documentation for at least a part of the activities which this sub-project was able to cover. There is, for example, no regulating framework covering in detail how the important safety measure "reduced speed" can be competently realised with the aid of a computer. Furthermore, highly important safety aspects such as in the design of the complete system are almost completely forgotten.

The operation of the industrial robot at full speed when a person is in the danger area of the robot is completely ignored or quite simply forbidden. It is this very mode of operation which is to be expected in the future to an increasing degree.

The information in the regulating framework concerning peripheral equipment and the construction of industrial robots is generally adequate for the basic problems. If, however, consideration is given to modern applications, no adequate information on the safe implementation of this modern technology is to be found. In particular, no reference is made as to how large areas of work, e.g., in the case of mobile robots, are to be protected.

The regulating framework gives only little information, and even less in detail, on the matter of ergonomics in the interaction between man and machine. We estimate that this point will become more significant in the future as the systems become increasingly complex.

Many documents are to be found under the heading "Structures and Principles for Safe Computer Systems". Hardly any connection is made, however, between the application of a certain structure and the resultant possibility of excluding faults as would be necessary for an assessment of safety. A list of all possible faults in a programmable electronic system was virtually nowhere to be found in the first place.

A very significant point - that of the quality and the accuracy of the software used in the robot systems - is given only inadequate treatment. Here it would seem that the necessary scientific basis is currently missing.

Safety requirements of a qualitative nature have been imposed on the hardware of programmable electronic systems up to now. These requirements amount to something like "use a higher-order safety device" or "use a redundant system". In actual fact virtually no demands have been made on the quality of the hardware, that is, it has practically been left up to the manufacturer whether he uses selected qualified or unqualified components in his control systems. It emerges that qualitative requirements e.g., break down probability of the system, must be incorporated into the regulating framework in the future. Certain quality levels for hardware should be clearly stipulated in this context.

ANALYSIS OF AREAS REQUIRING FUTURE WORK

In the main, the gaps and areas found in the regulating framework reflect the very same gaps and areas which were ascertained during the listing and assessment of safety aspects of industrial robots. In addition to the problem areas "mobile robots", "powerful tools", "avoiding collision" and "complex production environment" of the regulating framework other gaps became apparent which call for basic work in the technical and organisational areas as well as in the area of learning theory.

Generally speaking, a programmable electronic system possesses all relevant information on the status of the system. This information can be used to prevent accidents. This PES-specific procedure is already being used today in the safety measure "plausibility check" but it must be developed further. It thus ought to be possible, for example, for the programmable electronic system to prevent the starting-up of a linked production process if the operator accidentally actuated the sensor to start up the process when only wishing to remedy a production fault. The participants in the sub-project are of the opinion that the latent possibilities offered by the programmable systems have been exploited not nearly enough to date.

Of particular relevance from a safety aspect is the protection of a person when in the direct vicinity of the industrial robot. A few suggestions do exist as to the safety measures to be taken but work is still to be done, as, for example, in the case of reduced speed, on how this reduced speed is to be realised - also if a fault occurs in the system. The acceptance and practicability of many suggested safety measures must be rethought.

CONCLUSIONS AND RECOMMENDATIONS

The project participants were agreed that great progress could be made in the area of safety at work if it were possible to develop a really reliable and economical human body detector. Great efforts would be necessary in this area.

The sub-project 4 participants were of the opinion that, apart from the areas analysed which must be developed within an industrial robot context - completely new in some cases - there are also problem solutions available mostly from the large-scale production sector which could be applied to the safety problems of the industrial robot. The application of software tools in the development of robot software is to be mentioned in this context. We today still lack an easy-to-operate, generally applicable, and above all, affordable tool for this purpose.

We believe that the application of a test tool, a safety measure long known in the aviation industry, which is specially tailored to a certain robot type would considerably increase the safety of the operator during repair and maintenance works.

The use of a specification language would also constitute a considerable contribution to software safety and especially so if it were possible to develop this tool to such an extent that the actual program could be generated automatically from the specification.

As we know, perfect techniques are not alone decisive for the safety of a system but also, to a quite considerable degree, the behaviour of the person involved. There are many ways to teach a person to be conscious of personal safety. We consider the most important of these to be the training of employees and the ergonomic design of the place of work. This raises questions as to how a person receives information material or how a person is motivated to learn. These are all questions, by the way, which are being researched with the aid of artifial intelligence nowadays.

Whilst "ergonomics" and "questions on the theory of learning" cover relatively large complexes two closely-definable gaps have arisen from the case study. On the one hand, mention must be made of the lack of guidelines for particular applications. It is not enough to prepare a generalising regulating framework for the assessment of safety at work of an industrial robot installation but rather the relevance of a special application must be considered.

On the other hand, we were surprised to discover that apparently no thought had been given as to whether a programmer is in the position to push the emergency stop button on the manual programming unit should the industrial robot malfunction.

All this proves that there is still much to be done in the area of robotics safety.

GUIDELINE FRAMEWORK FOR THE ASSESSMENT OF PES

B. K. DANIELS

National Centre of Systems Reliability,
Wigshaw Lane, Culcheth,
Warrington, WA3 4NE, UK

INTRODUCTION

Ron Bell covered (1) the purpose and objectives of the CEC collaborative project. This paper, one of the collection of papers presenting the results of the collaboration, reports on objective 5.1, and the success of the project in devising a Guideline Framework for the Assessment of PES.

The framework was established taking into the account the data collection activities reported by Barry Stevens (2) and the results of the collection and assessment of current standards and guidelines for PES reported by Ole Andersen (3). Andersen reported in the recommendations of objective 3 the need to develop a top level document to contain a framework for the overall view of a PES and to assist the process of correlating and harmonising existing standards and guidelines.

This project result provides a framework which can be used in a wide range of PES applications and can co-exist and interact with many existing criteria for the safety performance of systems including PES. This paper concentrates on the framework, but Tony Wray describes (5) an example application to a robot system.

SCOPE OF THE ASSESSMENT FRAMEWORK

The framework provides for a systematic examination and assessment of programmable electronic systems used to provide safety related functions within much larger industrial systems. The framework is generic and can be applied to a wide range of PES applications, in most industries and accommodates existing safety criteria in use in the UK, France, the Federal Republic of Germany and Denmark. It thus differs in scope from all existing assessment guidelines which could apply to industrial applications of PES in Europe.

The framework is also sufficiently flexible to accommodate future guidelines, and encourages their development. Existing guidelines and knowledge are examined by Germer and Meffert (4) and found to be inadequate. The framework provides for these future developments and extensions, whilst immediately assisting the process of assessing the PES.

Ron Bell (1) reports the project proposals for the future development and review of the framework.

PARTICULAR ASPECTS OF PES SAFETY REQUIRING SPECIAL ATTENTION

PES based technology offers many advantages, including safety advantages, over other means of implementing safety systems. The use of PES is thus to be encouraged. But there are a number of safety aspects which need to be satisfactorily addressed.

A PES or its software may contain faults caused by errors in design. A PES can rarely be exhaustively tested under all possible conditions and so design errors may remain unrevealed for long periods of time. A particular set of operating conditions causes a program failure and reveals the underlying fault. Whilst testing is necessary in this context, it cannot be considered as sufficient.

A fault may be induced into the software or stored data as a result of some transient, a failure, or some disturbance. Means of detecting undesired changes to the contents of memory will be a desirable if not always required feature of a PES used in safety functions. Modern electronic devices are susceptible to electrical interference and special attention has to be paid to electro-magnetic compatibility (EMC).

The failure modes of PES are more complex than for earlier technologies. The rapid evolution of PES reduces the experience phase for a particular generation of equipment. PES are used where the ability to program is an attractive feature, thus making that PES a unique installation. The lower volumes of PES equipment type specific data and these other factors make the prediction of failure effects more difficult than for earlier technologies.

FRAMEWORK RATIONALE

The framework was developed through discussion over the two years of the technical programme of the collaborative project. It draws extensively on the experience of all the project members in assessing PES in their own industries and to the national and institutional requirements current in the 4 countries.

The nature of a PES has been defined in (1) and this paper will not repeat the basic definition. It is emphasised that the framework deals with one or more PES used in a safety related function, and that each PES may have one or more programmable elements.

The guidelines framework was developed:-

1. To enable PES, integrated into complex process and parts manufacturing and assembly systems, to be examined systematically in a 'top-down' manner.

2. Knowing that the technology and utilisation of PES is rapidly developing, it is essential that the guidelines framework avoids unnecessary restrictions on the detail design or use of PES.

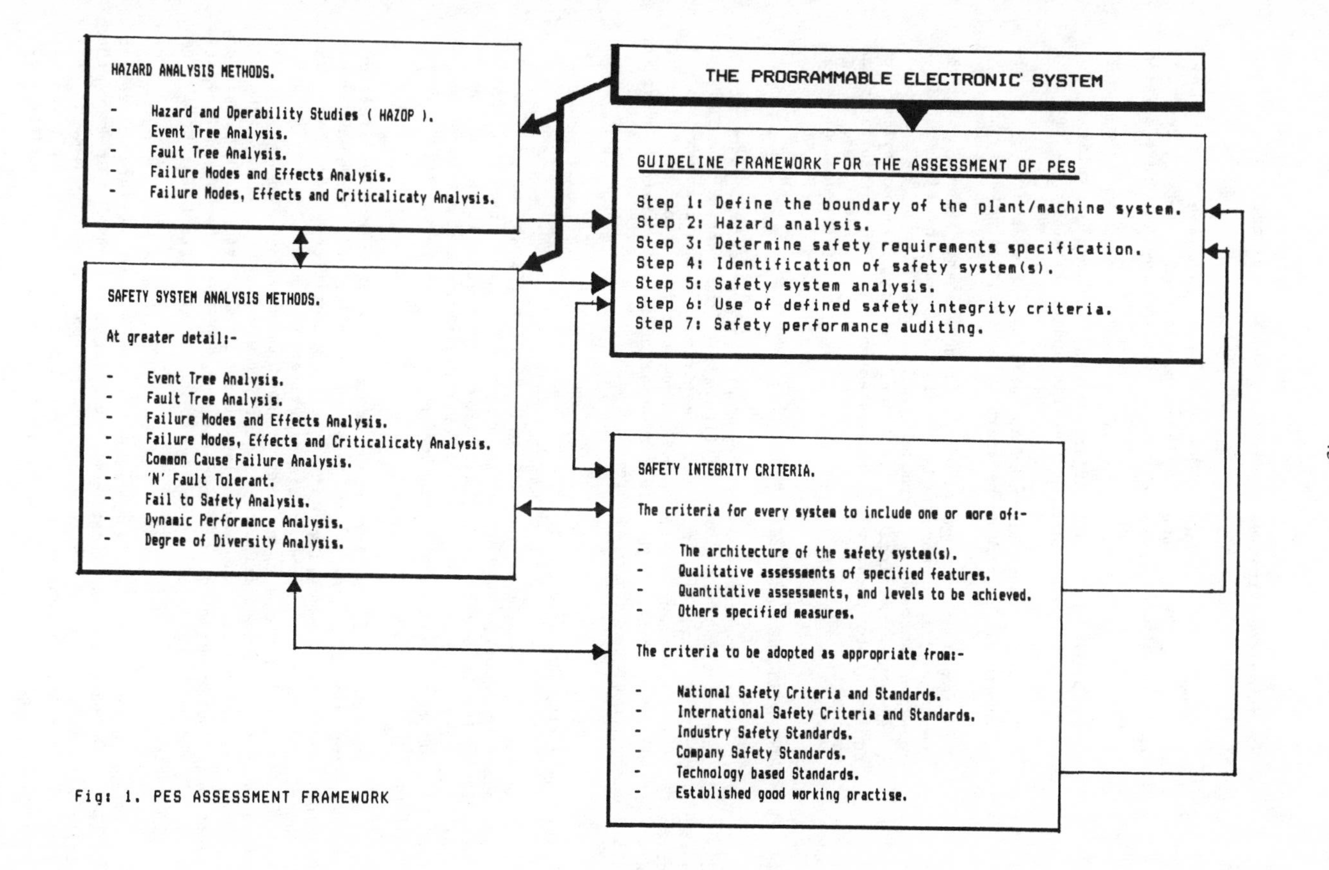

Fig: 1. PES ASSESSMENT FRAMEWORK

The guidelines framework should also be applicable to non-PES systems.

3. To take into account the various approaches that have historically been adopted to assess system safety integrity. The guidelines framework should therefore be particularly useful:-

   - To standards making organisations, to develop guidance for this area which is more specific and applications oriented.

   - To those persons who need to carry out assessments of systems incorporating PES where there is no existing guidance.

4. To allow a wide range of existing assessment techniques to be used, new techniques to be developed, and yet still retain a cohesive assessment methodology. This allows the type of assessment to include personnel safety, vulnerability of equipment to hazards and the assessment of PES reliability and availability.

THE FRAMEWORK STRUCTURE AND DETAILS

The framework comprises seven Steps which are described in detail in the following sections and shown in Figure 1. The framework calls for safety integrity criteria. It is not the purpose of the framework to set criteria, nor to comment on what are adequate criteria or levels to be achieved within criteria. The discussion on criteria therefore precedes the description of the framework Steps.

Safety Integrity Criteria

The setting of safety integrity criteria for PES is not part of the assessment framework, and so does not have a Step number. Safety integrity criteria are necessary to allow a judgement to be made as to whether a particular PES has met the criteria which apply to the use of the PES at that particular location and in performing the required functions.

In some countries, and for some industries, and for some applications of PES, and for some levels of hazards, and for some frequencies of hazards, and for some licensing/safety authorities, and for some industrial users there are established criteria. Sometimes the criteria may have a legal basis, or be based on good established practise, or may serve only as a recommendation. Frequently the criteria will not make any special provisions for the technology used in achieving safety related functions. But some criteria were derived for technologies which pre-date that of PES and are highly specific to the then current safety system technologies.

Thus it may be that a particular PES has to satisfy different criteria in the same application in varying locations. Or that the criteria that apply to a PES in a particular application may make it very difficult for PES to achieve the required level because the criteria are particularly adapted to enable earlier technologies to achieve safe operation. The design of a PES may have to be heavily influenced by the criteria that have been used for decades in a particular industry.

The assessor will need to establish which criteria have to be used in an assessment. The chosen criteria may then influence the detail safety system analysis which is carried out, since the analysis will need to provide the qualitative and quantitative results to compare against like in the criteria.

That there are a wide range of criteria that could apply to European PES applications was clearly established by the project. Examples are given in (5) and a longer list is available in (6). It was less obvious that there was a need for the existing variety of criteria. However the independent historical national, state and industrial development of law, standards, safety authorities and practices have led to the current divergent criteria. It must be recognised that changing and/or harmonising the criteria will be a lengthy process. Safety practitioners have not previously had to deal with the pervasive influence of a single technology such as PES to implement safety systems, and so have not had the need to establish a unified set of criteria which could be applied in all European Community Countries..

STEP 1: Define the boundary of the plant/machine system

The purpose of this step is to encourage the adoption of a systematic approach to the assessment of a machine or plant system where safety is to some degree dependent upon the correct functioning of a PES. There is a need to consider who and what is put at risk by the operation of the plant or machine to be assessed.

There may be several groups who are at risk, and the risk may vary with the machine or plant operation and over time. The boundary could be dependent on a number of such factors and may need to be redefined when Step 2 has identified and analysed the hazards.

An important aspect of this assessment framework methodology is the need to take into account information gained in later Steps affecting consequentially decisions made in former Steps. This will frequently necessitate iteration around the Steps until no further consequential changes are identified.

The Step 1 will need to record the decision process towards a boundary, and the exact boundary and any discussion on alternatives considered, modified, or rejected. This recording process is a further important feature of the methodology since the results of the assessment will only be valid relative to the chosen boundary.

Assessing the safety of the plant or machine when it is subject to proposed modifications, changes of use or environment, or to meet different safety requirements, or through changes in the Step 1 boundaries of neighbouring systems, will require the full 7 Steps of the framework to be reconsidered. Re-use of an earlier assessment as the basis for the new assessment is expected, and where necessary this must be reworked to a new boundary and to include the new conditions.

STEP 2: Hazard analysis

There is a need to identify the hazards which could arise in all modes of system operation and to determine the events leading to those hazards.

For each identified hazard, the level of severity/consequence should be determined. For personnel (and equipment) an estimate should be made of the frequency and duration of exposure to the hazard.

Hazard analysis may be informal for some simple manufacturing systems and be based on the collective experience of safe design and operation of such simple systems. Where the system is more complex, more rigorous methods must be used including:-

- Hazard and Operability Studies (HAZOP).
- Event Tree Analysis.
- Fault Tree Analysis.
- Failure Modes and Effects Analysis (FMEA), and its variants including Failure Modes, Effects and Criticality Analysis (FMECA).

The hazard analysis may have already been carried out for a similar system or for another safety assessment on this system. In these cases the assessor should carefully check to ensure that the current application matches the application for which a hazard analysis already exists. If there is a close correspondence then the existing analysis may be used. Where there are differences, a new (full or partial) analysis will be necessary.

STEP 3: Determine safety requirements specification

The correct and complete safety related functions that the total system must perform in the circumstances covered by the system boundary and hazard analysis Steps must be specified. This specification can then be used as the basis for the safety assessment of the system.

It may be that the safety requirements specification has already been developed for a similar system or for another safety assessment on this system. In these cases the assessor should carefully check to ensure that the current application matches the application for which a safety requirements specification already exists. If there is a close correspondence then the existing analysis may be used. Where there are differences, a new (full or partial) re-definition of the safety requirements specification will be necessary.

STEP 4: Identification of safety system(s)

The total system within the exact boundary may have parts which have safety related functions and others which have no such functions, and some which provide both safety and non-safety functions. Figures 2(a), 2(b) and 2(c) illustrate this consideration. Normally the non-safety functions of a system do not require a detailed analysis as in the next Step. However some non-safety functions may themselves lead to safety related events in the event of failure or particular operational circumstances, in

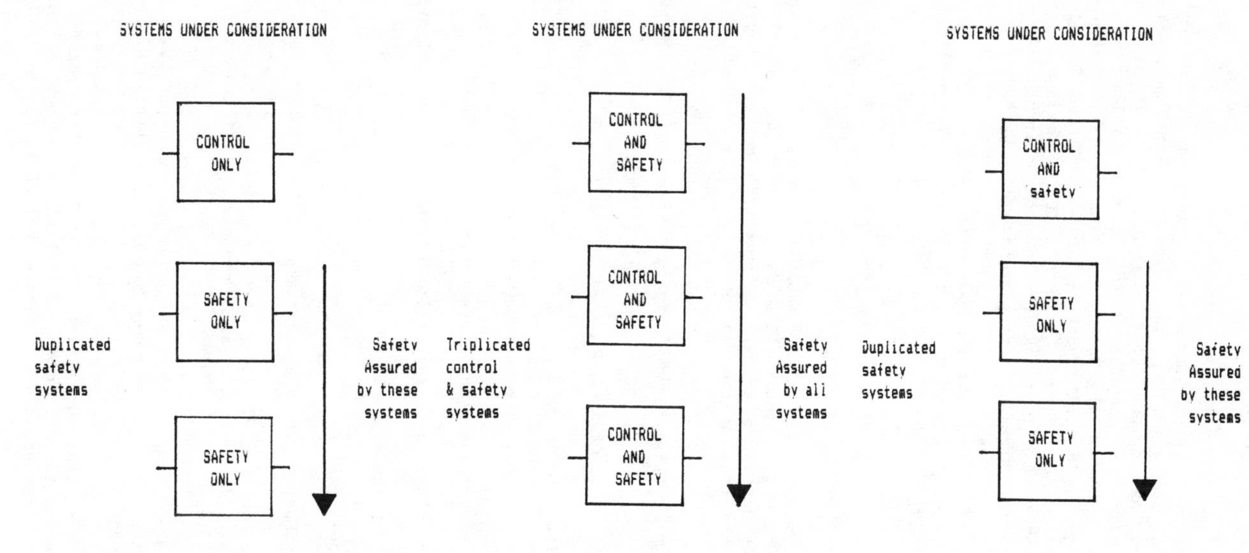

Safety functions and control functions separated. Providing the safety systems cover all dangerous failures, taking into account the likely demand on the safety systems, then the assessment process can be directed at the dedicated safety systems.

Fig: 2(a).

Safety functions andcontrol functions combined. Assessment process must cover all systems.

Fig: 2(b).

This configuration is essentially the same as figure 2(a) but additional non-essential 'safety' functions have been integrated with the control system. Non-essential safety features cannot form part of the Safety Case for the PES, plant or machine. Treat as figure 2(a).

Fig: 2(c).

cases they should be subject to detailed analysis to scope the effect on safety system behaviour.

The part(s) of the system which have a responsibility for safety should be identified. These safety related systems should have a function to recognise an event or sequence of events that may lead to one or more hazards, and to curtail the events leading to a specific hazard or to mitigate the consequences of the hazardous events. These events will include those identified in earlier Steps, but should also include those which are identified in later Steps. This will further call for iteration through the framework Steps until no further consequential effects are apparent on the boundary, hazards, safety requirements or safety systems.

Identification of the safety systems may also lead to savings in effort in later Steps by allowing identical parts of parallel safety channels to be subjected to a single channel generic safety system analysis and comparison with safety criteria.

STEP 5: Safety system analysis

This step allows a variety of methodologies to be used to assess the safety systems identified in Step 4. The chosen method(s) and the breadth and detail use will depend on the safety integrity criteria adopted and are likely to be the ones listed for Step 2, but conducted at a more detailed level.

Examples where criteria could influence choice of methodology include:-

1.1 The architecture of an individual safety system.

1.2 The architecture of multiple safety systems.

1.3 The degree of diversity in hardware and software.

2. Qualitative analyses (and tests) which could cover:-

2.1 Errors or omissions in the safety requirements specification.

2.2 Systematic hardware failures.

2.3 Common cause failures.

2.4 FMEA at component level a requirement.

2.5 Operation of the PES with faults injected (hardware and/or software). These could allow comparison with criteria such as:-

- The maintenance of safety performance in the presence of N failures, where N is prescribed in the criteria (e.g., single component).

- Fail-safe behaviour. Where the failure of components are as defined in the criteria, and the safe behaviour of the system must be demonstrated in the presence of those failure modes.

- Dynamic performance. Where the dynamic characteristics of the PES can determine the safe behaviour.

3. Quantitative assessment of the random hardware failure functions (e.g., dangerous failure rate, probability of failure on demand, spurious safety action rate). The figures obtained could be compared with targets in the criteria, or as a basis for comparison with comparable situations (PES replacing a non-PES safety function, PES design alternatives).

4. Other specific measures, for example:-

   - Organisational structures for design, implementation, maintenance, test etc.

   - Established performance, for example requiring a certain number of years experience with a PES in non-safety use before use to provide safety functions.

   - Type assessment of the PES at component or system level, and thus re-usable assessments.

   - Quality assurance programmes.

STEP 6: Use of defined safety integrity criteria

Having assessed the PES in such a way that the results can be compared with the adopted criteria, this Step calls for the comparison to be made. It is important to clearly distinguish between the three stages of setting and adoption of criteria, assessing the PES, and comparing the PES assessed safety performance with the criteria.

The assessor should carefully check that the adopted criteria are relevant to the PES application. Although this framework principally addresses the safety of personnel, the criteria may also call for standards to be met in the protection of equipment and the product quality. It is likely that in the majority of industrial applications of PES that a set of criteria will have to be satisfied.

The comparison may show that some criteria are met, whilst others are not. It may be that there are priorities within the set of criteria, such as personnel related criteria being compulsory whilst equipment criteria are advisory. It will be necessary to resolve any conflicts. This may call for design alternatives to be studied. The assessor should ensure that any changes proposed are carefully examined for their potential consequences through all the Steps of the framework. If there is doubt over a critical criteria, then a complete rework of the whole framework may be justified taking into account the sensitivity of the result to the various decisions and assumptions.

STEP 7: Safety performance auditing

The Steps 1 to 6 will usually be applied to new systems but Step 7 is applicable to an already assessed system during it useful life.

The need to carry out a safety audit may be due to planned changes to the system to improve performance after taking account of performance records, meet new safety requirements, or to carry out new functions. Some systems may have a safety audit conducted periodically in order to meet the criteria or licensing procedure.

The purpose of the audit is to show that:-

- Records of system performance, design, implementation, and corrective maintenance are accurate.
- The operational performance of the PES is in conformance with the criteria.
- Where no deliberate changes have been made to the PES, that the PES is exactly the same as when the original assessment was performed.
- Where planned changes have been made, that the records of the system are updated, and that the PES is as expected from the records.

The disciplined approach to assessment must be continued into the maintenance phase and until the system is decommissioned to secure the PES against deterioration.

CONCLUSIONS

The collaborative project has met the objective 5.1 and provided a guidelines framework for the safety assessment of PES in industrial applications. The framework is generic and may be used with previously existing safety criteria and well established assessment techniques. It avoids the need to freeze the PES technology at the current state of the art and practise. It may be extended and related to new technologies and safety standards as they are developed and brought into use.

REFERENCES

1. Bell, R., Assessment architecture and performance of Industrial Progrmamable Electronic Systems (PES) with particular reference to Robotics. See this Symposium.

2. Stevens, B., Presentation of Objectives 1 & 2 of the Joint Collaborative Project on Programmable Electronic Systems Collection & Data Banking of Information. See this Symposium.

3. Andersen, O., Collection and assessment of current standards and guidelines for PES - CEC Collaborative Project. See this Symposium.

4. Germer, J., Meffert, K., The inadequacies of research into programmable electronic systems in industrial robots. See this Symposium.

5. Wray, A. M., Case study using the Guidelines Framework. See this Symposium.

6. CEC Collaborative Project, Assessment, Architecture and Performance of Industrial Programmable Electronic Systems with particular reference to robotics safety, Final Report, 1986.

CASE STUDY USING THE GUIDELINES FRAMEWORK

A M WRAY

Safety Engineering Laboratory
Research and Laboratory Services Division
Health and Safety Executive
Sheffield
UK

ABSTRACT

A collaborative project, involving organisations from the UK, W Germany, France and Denmark, funded by the EEC and aimed towards the harmonisation of European guidelines for Programmable Electronic Systems used in safety related applications, was concluded in 1985. Objective 5 of this project developed a Guidelines Framework for the application of existing Safety Integrity Criteria in a standardised stepwise manner. This paper describes the application of this Guidelines Framework in the limited safety assessment of the programming mode of a robot installation and is intended only to illustrate the use of the Guidelines Framework.

INTRODUCTION

The Guidelines Framework, developed as part of Objective 5.1 of the EC funded Collaborative Project (references 1 and 2), enables Programmable Electronic Systems for use in safety related applicaltions to be assessed in a stepwise manner using existing Safety Integrity Criteria. This limited assessment of a robot installation is intended to illustrate the use of the Guidelines Framework and not be a specific application of Safety Integrity Criteria which will vary from country to country. The installation was identified as part of the Collaborative Project and a complete description is given in reference 2. Only those aspects associated with the PROGRAMMING MODE of the robot are considered.

DESCRIPTION OF THE SYSTEM UNDER CONSIDERATION

The system is for semi-automatic arc welding and consists of two stations which are used alternately by a central welding robot; while work is being loaded/unloaded from one station the robot is welding at the other (Figure 1). All control functions (eg sliding door operation, control of the rotating tables etc) are carried out by the robot control system.

The majority of the instructions for controlling the robot and other equipment such as the rotating tables, access door latches and the response to limit switches and the "two hand" controls etc, is programmed

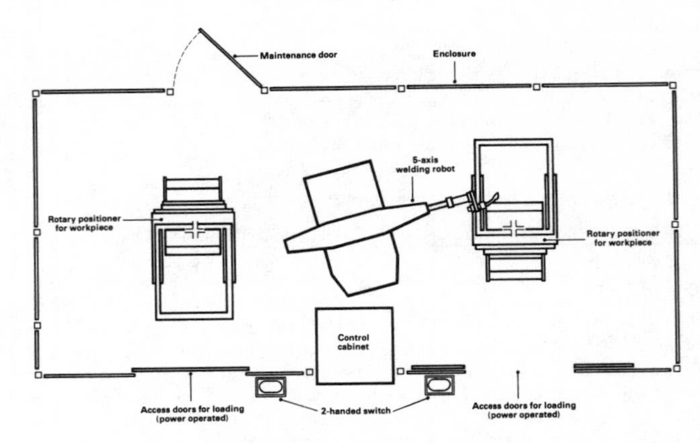

Figure 1 - The robot installation under consideration

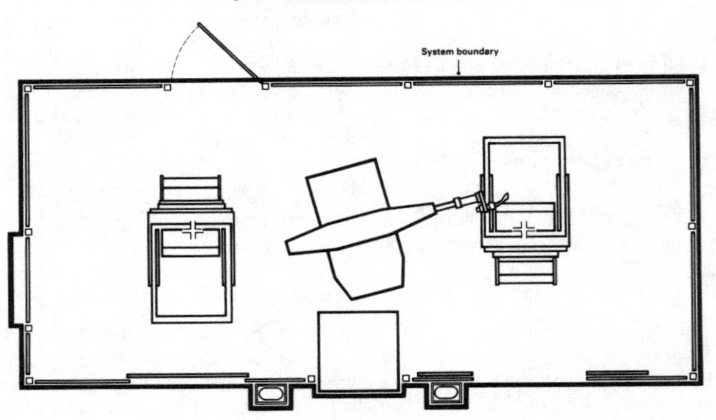

Figure 2 - Showing the initial system boundary

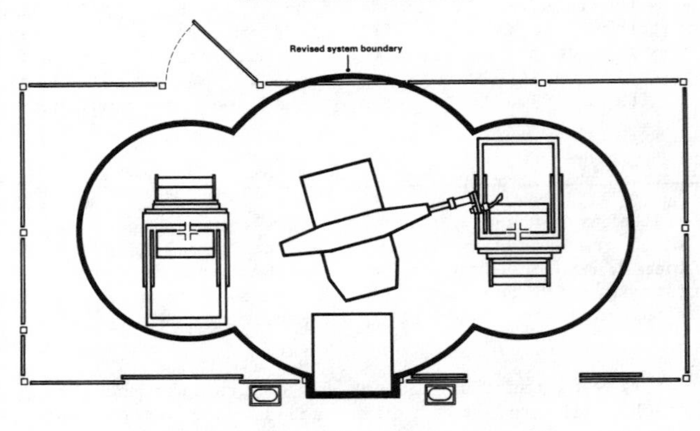

Figure 3 - Showing the revised system boundary

into the controller using a visual display unit (VDU) as the programming terminal. The remaining instructions relating to the exact positioning of the robot, require close observation by the programmer and these instructions (or their positional coordinates) are set into the program using the programming pendant. It is the use of this pendant which is specifically considered in this case study.

STEP 1: DEFINE THE BOUNDARY OF THE PLANT/MACHINE SYSTEM

The boundary to be considered in any assessment depends on the activities that the personnel involved are to perform, together with the hazard being considered. At this, the first, stage in this assessment the boundary must encompass all of the possible hazards associated with programming the robot but it excludes the potential hazards associated with the crane used to load and unload the workpieces as this is considered to be a separate, independent but overlapping problem. The boundary will, however, need to be reassessed and revised as the study progresses and further information is obtained. Figure 2 shows the boundary which was initially considered and, in this case, is the physical boundary coincident with the fence surrounding the installation.

For the purpose of this study, in which we are concerned only with the hazards to the programmer associated with the operation of the robot in its programming mode, the boundary may be revised as follows. During programming all of the interlocking functions associated with the perimeter fence are inoperative. This is because the robot controller is used to carry out all PES functions within the installation. This includes the door interlocking which is consequently under the direct control of the programmer. Therefore, during programming, access is available to both welding stations simultaneously and so the boundary constitutes the envelope of the robot workspace and the rotating tables in both workstations regardless of which one the robot is being programmed to service (see Figure 3).

It should be noted that if the fence is of a light construction and the robot is capable of penetrating it (ie there are no mechanical stops on the robot) then the boundary must include that area of the robot workspace which projects through the fence.

The installation being considered has a welding head permanently fixed to the robot arm as opposed to a gripper. If a gripper had been fitted it would have been necessary to consider the possibility of workpiece ejection and would require a widening of the boundary to include the maximum range of any ejected part.

STEP 2: HAZARD ANALYSIS

The purpose of this Step is to identify the hazards and to determine the events leading to those hazards.

The potential hazards during the programming operation are those of impact from the robot and crushing by the robot against stationary objects or scissor action at the joints. In this example either impact or crushing could be very severe.

The robot controller controls the rotating tables on which the workpieces are mounted. Contact with one of these or the workpiece mounted on it may similarly cause serious injury.

The total time of exposure of the operator to risk depends on many factors including:-

- The time taken to carry out the programming (including the testing of the program). This will depend upon the complexity of the task being programmed.

- The speed and programming ability of the operator.

- The frequency that (re)programming is required. This will depend on the product lifetime and the number of changes made to the product during that life.

The task in the case study (welding of tractor mudguards) is presumed to have a relatively long lifetime (years) and has an average programming time (in which the programmer is within the robot envelope) of about 4 hours for the first mudguard, although some reprogramming may be necessary during the production run. This aspect of risk may need to be considered in greater depth depending on the safety integrity criteria to be adopted.

In view of the very limited number of discrete hazards that are present in this installation it was not considered necessary to adopt any complex techniques of hazard analysis in this Step.

IDENTIFICATION OF THE HAZARDS

The hazards relevant to this assessment, which is concerned solely with the programming mode of the robot, are:-

1 Impact from the robot, including crushing against stationary objects and scissor action at the joints or

2 Impact from the rotating tables or the loaded mudguard.

3 Other minor hazards eg burning from the hot welding tool, eye damage caused by emissions from the arc, etc

Only item 1 will be dealt with in the study and, for the purpose of this assessment, any inadvertent contact with the moving robot constitutes a hazard.

DETERMINATION OF THE EVENTS LEADING TO THE HAZARD

Figure 4 is a basic event tree showing the interrelationships between the various contributing factors to the above hazards which will now be described. Note that the area with which the guidelines framework is primarily concerned is shown dotted in this Figure.

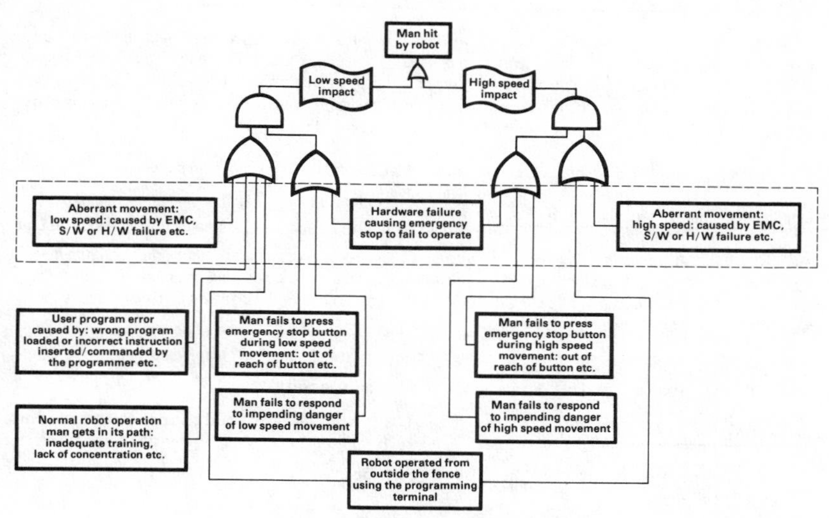

Figure 4 - Event tree considered in the assessment

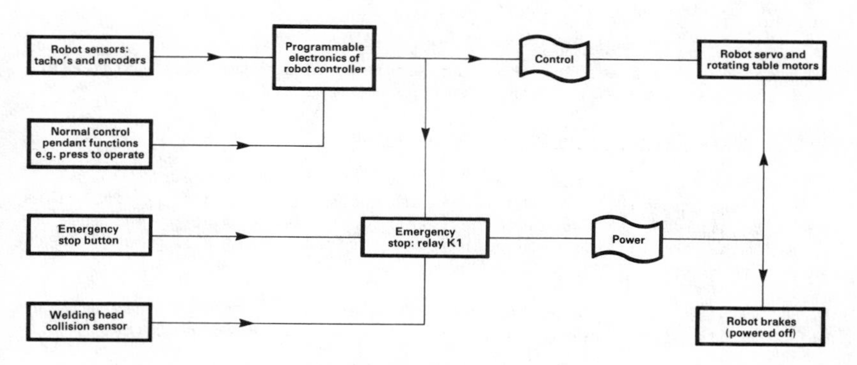

Figure 5 - Relationship between safety systems

Robot performs correctly but the operator gets in its path

This event is difficult to avoid in the system described except by a high degree of care on the part of the programmer. The use of the restricted programming speed, an emergency stop and an anti-collision switch around the welding head will reduce the probability of this event maturing into an accident involving operator injury.

Only low speed movement is associated with this event.

Part program/programmer error

Whilst stepping through a program, the programmer may be injured if the robot makes an unexpected movement. The unexpected movement may be due to a programming error, or the loading of an incorrect program for amendment by the use of the wrong cassette tape etc.

Because the speed limitation system is working correctly for this mode of failure only low speed movement can occur.

The robot moves aberrantly

Aberrant movement of the robot can be the result of many influences. For example:-

- Electrical interference or

- Robot control system failure.

Depending on the cause of the failure the robot may move at low or high speed. Two boxes are shown on Figure 4 associated with the failure giving either high or low speed aberrant movement.

Robot control from the outside of the fence

The operation of the robot from the outside of the fence by unauthorised manual operation is assumed to have to be deliberate. This would require the keyswitch to be reset to the auto position and the programming terminal to be used before a program could be executed in the automatic mode.

The site safety rules instruct the operator to take the key with him whilst he is within the fence in order to avoid this possibility.

Movement can be at low or high speed depending on the commands given to the robot.

Additional events contributing to the hazard

In addition to the above initiating events occurring, the emergency stop system must also fail to operate in order for the top event to be realized. The event tree in Figure 4 shows how these additional events relate to the initiating events described above.

With respect to Figure 4 it can be seen that there are two boxes labelled "man fails to press emergency stop .....". This is because the probability of the man failing to press the emergency stop button will depend on the time he has available to reach it which will, in turn, depend on the speed at which the robot is moving. This also applies to the two boxes labelled "man fails to respond to impending danger .....".

STEP 3: DETERMINE SAFETY REQUIREMENTS SPECIFICATION

The purpose of this step is to determine the safety requirements specification for the system under consideration which should be examined in the light of the hazard analysis carried out in Step 2. (It should be noted that a superficial analysis of the safety requirements specification should be done in this Step but only so far as to allow step 4 to be carried out; the full assessment being carried out in step 5.)

A definite specification was not obtainable for the system being considered but, as this study is only a limited safety assessment for demonstration purposes and information was available to enable the subsequent steps to be completed, the study was continued.

STEP 4: IDENTIFICATION OF THE SAFETY RELATED SYSTEMS

Figure 5 shows a diagrammatic representation of the safety related systems which are considered relevant in this assessment.

In order for any safety related system to be fully independent, it must be capable of sensing the impending hazard, recognizing the significance of this hazard and performing some function to mitigate or avoid the realization of the hazard independently of all other safety related systems. (Interlinking between the systems in order to provide monitoring of one system by another for example does not affect this independence providing the control function path does not go via any part of the second system.)

The safety related systems shown in Figure 5, and more fully described in the collaborative project (reference 2) report: Objective 5, are as follows:-

ROBOT CONTROLLER

Correct operation of the controller at all times is required in order to avoid any aberrant robot movement.

Speed limitation

The main function of the speed limitation of the robot whilst in the programming mode is to protect the programmer from his own errors. In the programming mode the speed of any robot axis is limited to about 25% of the maximum speed, under the control of software within the robot controller, as follows:-

- In the "continuous path" (CP) mode the robot is allowed to operate at its programmed speed (ie the speed set by the programmer) but subject to an upper limit of 500 mm/sec (20% of the maximum speed). Normal welding speeds seldom exceed 33 mm/sec so that, apart from a programming or system error, the likely maximum speed will be much less than this upper limit.

- In the "point to point" (PTP) mode the maximum robot speed is set to 750 mm/sec (25% of the maximum speed).

EMERGENCY STOP SYSTEM

This is the ultimate safety related system and, in general, will be used only when other systems (or practices) have failed. The emergency stop system is basically a loop which, if broken, causes a latching relay (referred to as relay K1) to drop out. This relay controls:-

- robot servo power

- power to keep the robot brakes in the OFF position

- AC power to the motors driving the rotating tables.

Thus when the relay drops out, the power to the robot and rotating tables is removed and the brakes are applied to the robot by hardwired means.

Emergency stop buttons

Several emergency stop buttons are provided including one on the programming pendant and another on the control panel which is accessible from the outside of the enclosure. They operate via the hardwired emergency stop system.

Collision emergency stop

A collision switch (termed a collision emergency stop switch by the robot system manufacturer) is mounted around the welding head which will stop the robot when a collision between another object and the welding head is sensed. It operates via the hardwired emergency stop system.

Robot brakes

The brakes on the robot, which can be seen from Figure 5, to form part of the emergency stop system, are applied by means of springs and are held off by electrical power. They are applied:-

- when the emergency stop system is operated or

- when mains power is removed.

STEP 5: SAFETY RELATED SYSTEM ANALYSIS

In this Step analyses of both the requirements specification and the safety related systems have to be carried out within the context of the safety integrity criteria to be applied in Step 6. The analysis will cover:-

- the safety requirements specification,

- the safety related systems to ensure that they fulfill the conditions and functions set out in the safety requirements specification, and meet the safety integrity criteria.

NOTE

It should be noted that Steps 5 and 6 must be carried out together as the form of analysis to be adopted in Step 5 depends on the criteria to be later applied in Step 6 (eg quantitative safety integrity criteria require a quantitative assessment of the parts of the system to which they apply). Conversely the safety integrity criteria cannot be applied until after the analysis has been carried out. Steps 5 and 6 are treated as being sequential in this study in order to provide a better explanation for the reader.

A full safety requirements specification which applies to the installation was not available so that an analysis of it could not be made and so this stage has been omitted from this study. It should be noted however that the design of the safety system rests heavily on the safety requirements specification and, indeed, it is tested against it. Any errors or omissions in the specification will therefore be reflected in the safety system. Although an analysis of the safety system specification could not be carried out in this example it should be remembered that this is a very important stage in a full assessment and cannot be neglected.

In considering the overall installation it can be seen that there is no automatic protection for the operator. (Although there is a collision sensor around the welding head, this does not protect the operator from the robot arm.) The severity of the possible injuries caused by the robot indicate that the installation is within a "serious hazard" category which therefore affects the safety integrity criteria to be used. However, in this case, although there is a separate and independent hardwired safety related system it is dependent on the operator for its operation. In addition the ability of the operator to react and operate the emergency stop button in time to avoid an accident depends upon the speed of the robot movement and hence on the robot failure mode. Thus a specialist judgement needs to be taken in forming a decision regarding the assessment in a case of this kind. It is clear that whilst generic safety integrity criteria are available for PES's, criteria for specific applications, such as this, which will take into account the requirements of the particular application, need to be developed. For example the criteria depend on the safety integrity associated with:-

1 the PES,

2 the hardwired emergency stop system and

3 the working practices adopted.

Note

The safety integrity criteria currently being developed for PES applications are generic, in that they are intended to apply to all PES based equipment and applications. Assessing the safeguards adopted in the case of an industrial robot application requires criteria that have been specifically tailored for such an application. However, the generically based criteria can be used as a basis for decision making.

Two safety related systems have been identified in the context of this case study (see Step 4) which would normally be considered adequate for the hazards present. However, one of the safety related systems, the emergency stop system, requires the man to operate it in the event of either his error or a failure of the robot controller. The emergency stop system thus relies on the man having time to react quickly enough to stop the robot before he is injured. This is the reason for the robot speed limitation in the programming mode. In the event of a failure of the robot control PES, there is a possibility that this could result in high speed movement of the robot and the question remains as to whether the man could react in time. (This needs considering in the context of programmer training and his skill in keeping outside of the robot envelope [or perhaps his desire to].) If the answer to this question is yes then it could be argued that there are 2 systems protecting the man (ie PES + hardwired backup). If on the other hand it is felt that for some proportion of PES malfunctions the man could not react quickly enough one could argue that there is only one safety related system protecting the operator (ie just the PES).

The following are considered:-

- Case (a): Only one safety related system is available ie that comprised of the PES.

- Case (b): Two safety related systems are available ie the PES and a hardwired backup.

STEP 6: USE OF DEFINED SAFETY INTEGRITY CRITERIA

Because of the differing approaches involved in the safety integrity criteria of the member organizations examples showing the application of the Berufsgenschaftlitches Institut fur Arbeitssicherheit (BIA) and the Health and Safety Executive (HSE) safety integrity criteria will be considered. The Guidelines Framework (references 1 and 2) or the respective organisations should be consulted if further details of the criteria are required.

APPLICATION OF BIA SAFETY INTEGRITY CRITERIA

The following table shows how the two cases described above, may be related to the BIA safety integrity criteria.

BIA system	Case (a)	Case (b)
System 1	Criteria not met: System 1 applies only to very low level injuries (no longer than 3 days absence from work). For the robot installation in question, very severe, perhaps fatal, injuries could be inflicted.	Criteria not met because the injury level may be high and only one safety related system is employed in BIA system 1.
System 2	Criteria not met: System 2 may be used only if it is possible to "avoid" the accident. This possibility is specifically excluded from Case (a).	Criteria met if it is possible to "avoid the accident". "Avoidance" of the accident could mean moving out of the robots path, operating an unspecified emergency stop device, etc.
System 3	Criteria not met: The hardwired back-up safety related system is not considered to be available in Case (a).	Criteria met only if it is acceptable that the programmer has to operate the non-PES safety related system (emergency stop).

APPLICATION OF HSE SAFETY INTEGRITY CRITERIA

If no specific safety integrity criteria apply for the application under consideration then the aim should be to ensure that the safety integrity achieved is not inferior to that which would have been achieved in similar situations using a non PES based system (ie similar level of hazards, frequency of access etc). In this case no specific safety integrity criteria is available and so a comparison of this type with similar situations is necessary.

It is assumed that the level of hazard is relatively high and the frequency of access relatively low so that two safety related systems would be required. The most appropriate safety integrity criteria to apply are contained in the following Table extracted from reference 2.

HSE system	Case (a)	Case (b)
Case 1	Architecture of Safety related system does not meet criteria - only 1 system is available.	Safety related system architecture suitable - 2 systems available. The Reliability and Quality requirements will also have to be met (see Table 2 of reference 2) and reference 3.

Once again a great deal depends on the ability of the programmer being able to operate the emergency stop system and hence on the probability of the system failing in a way which gives aberrant movement. Attention will need to be given to an analysis of the failure modes of the PES as well as other, less obvious, safety features such as operator training.

STEP 7: SAFETY PERFORMANCE AUDITING

Safety performance auditing is the practice of checking that a system is performing satisfactorily after its commission and ensuring that any subsequent changes to the system have been made and tested with suitable thoroughness. In addition to helping to ensure the continued safety of the system in question, this practice also provides feedback to the assessor regarding his original assessment.

As far as this example assessment is concerned it is not possible to audit the performance of, nor the changes made to the system.

CONSIDERATION OF FURTHER ITEMS FOR INCLUSION IN THE GUIDELINES

Whilst it is appreciated that there is a need for generic safety integrity criteria to be developed, there is also a need to develop application specific criteria which take into account the particular requirements of the different types of equipment. This should, of course, be carried out on a European, rather than national basis in order to develop European standards and so facilitate the interchange of technology.

REFERENCES

1 Guideline Framework for the assessment of PES, B K Daniels, this symposium.

2 Final report of the EC Collaborative Project Assessment, Architecture and Performance with particular reference to robotics safety: 1986.

3 Guidance on the use of Programmable Electronic Systems in safety related applications; Part 1: General requirements, Health and Safety Executive, UK, draft document for consultation, 1985.

USE OF PROGRAMMABLE ELECTRONIC SYSTEMS IN INDIAN NUCLEAR POWER PLANTS

S.N.AHMAD, U.N.PANDEY, K.NATARAJAN

Nuclear Power Board
(Department of Atomic Energy)
Homi Bhabha Road, Colaba
Bombay-400 005, India

ABSTRACT

The use of programmable electronic systems is finding increasing use in Indian nuclear power plants. This paper reviews the evolution of the control & instrumentation system in Indian pressurized heavy water reactors and highlights the areas where programmable electronic and computer based systems are used as well as details the plans for the future.

OBJECTIVES OF THE CONTROL SYSTEM

The primary objective of the control system is to ensure safe and reliable operation of the nuclear power plant, as well as its safe shutdown even during upset conditions (e.g. during a seismic disturbance). Since the power ramp in a nuclear reactor can be very steep, fast acting control systems are required for these reactors. Should any of the parameters exceed the permissible limits, the protective system would be required to shutdown the plant and keep it in a safe shutdown condition. The protective system has to be ultra reliable and capable of measurement and control over a wide range,and fail-safe in operation.

Economy dictates that nuclear power plants have high availability. This trade off between safety, reliability and availability has been achieved by making the control and protective systems in,critical reactor systems,triplicated. A two out of three coincidence causes a reactor trip. This scheme permits any one channel to be taken out of service should it develop a fault, or for testing,and still permit the plant to be operated safely.

In this paper, the discussion will be restricted to the Pressurized Heavy Water Reactors (PHWRs) in India.

VARIETY OF CONTROL SYSTEMS

A nuclear power plant consists of a large number of inter-related systems that require to be controlled precisely. The control system of the nuclear power plant is therefore large and complex. The major systems in a typical PHWR are listed below:

Reactor Protective System
Reactor Regulation System
Primary Heat Transport System
Moderator System
Channel Temperature & Flow Monitoring System
Shield Cooling System
Ventilation System
Boiler Water and Steam System
Turbo Generator System
Electrical System
Dousing Water System/Suppression Pool System
Fuel Handling System
Annunciation System
Auxiliaries like the Fire Detection & Fighting System,
Leakage Detection, Lekage Collection, Sampling System.
Radiation Monitoring System (including area monitoring & detection of activity in air/water & gaseous discharges).
D2O Upgrading Plants

The large number of systems include a variety of equipments and processes. The control systems for these systems of necessity involve a wide variety of controls using a multitude of control strategies and work relating to a number of disciplines. Typically, this relates to work in the disciplines of electrical and electronic control systems, computer control, process control, fluid power controls (involving several fluids like heavy water, oil, light water & air) and mechanical systems.

A large centralized control room has been provided for remote control of the various processes of the nuclear power plant.

EFFORT AT INDIGENIZATION

From its every inception, the nuclear power programme in India has placed great emphasis on increasing indigenization. This has been particularly true of the field of control and instrumentation. Even for Tarapur, which was the first nuclear power plant to be set up in India, a major part of the control and instrumentation had been built in India. The indigenous content for control and instrumentation has progressively increased in Rajasthan Atomic Power Station and later the Madras Atomic Power Station. The Electronics Corporation of India Ltd., an undertaking under the Department of Atomic Energy, has been a major contributor in this effort. For Narora Atomic Power Project, this trend has been temporarily reversed because of the higher level of computerization as well as the higher level of qualification demanded of critical items (e.g. seismic qualification), use of MIL qualified equipment etc.

CONTROL & INSTRUMENTATION (C&I) SYSTEMS OF RAJASTHAN ATOMIC POWER STATION

The Rajasthan Atomic Power Plant was the first PHWR power plant set up in the country. The control room panels in Rajasthan contain all the major displays and controls essential for normal and emergency operation of the plant. The relay based annunciation system, draws the operator's attention audio-visually to alarm conditions. Most control of processes is done using electronic PID controllers and transducers that convert the process signal into electrical current signal. The main plant logic system and the plant protective system are relay based. The reactor regulating system is electronic based, using SSI ICs. An electronic data logging system had also been originally provided. The fuel handling system uses discrete component RTL logic. Some upgrading has been done for some C&I Systems in Rajasthan. The major areas are:

a) Provision of a process disturbance analyser system
b) Microprocessor based failed fuel detection systemm
c) Computer based channel temperature monitoring system

C & I SYSTEMS IN MADRAS ATOMIC POWER STATION

The C & I systems in Madras are similar to Rajasthan except that the indigenous content has been increased and there has been upgradation of some systems. A more well defined and stringent Quality Policy has also been followed in Madras. All this has resulted in marked improvement in the performance of the C & I systems in Madras.

Among the major upgradation done in Madras are the following:

(a) On-Power Fuel Handling System

The Ge transistor based RTL system has been replaced by Si transistor based DTL system. Improved design has been done for trip units, certain logic cards, 2220 hz power supplies, high current dc power supplies and time delay units. Upgraded components have been selected and used, higher derating factors have been followed and all round quality has been improved.

(b) Channel Temperature Monitoring

The channel temperature monitoring system in Rajasthan had problems with core memory, which required controlled atmosphere. In Madras the system has been designed round a microprocessor using solid state memory. Since the system reliability was better, the analog indication used in Rajasthan were removed and system display provided, through microcomputer, on a CRT. Control room annunciation for alarms is given through the processor. Alarm hard copy makes location of channel causing alarm easier.

CONTROL & INSTRUMENTATION SYSTEMS FOR NARORA ATOMIC POWER STATION

Substantial improvements have been done in Narora C & I systems over the earlier projects. One area is increased computerization, where computer based control systems have been used for on-power fuel handling controls, reactor controls, plant data acquisition system and other critical areas of the plant. Some of these are described in detail below:

(a) On-Power Fuel Handling Control System

On-Power fuelling capability in PHWRs gives these plants many advantages, like better utilization of fissile material, lower system contamination and better availability/capacity factors by avoiding long and costly refuelling outages. However, this requires a complex fuel handling system and a sophisticated control system for the same.

The fuel handling control system consists of the following major subsystems:

- Computer system
- Electrical/Electronic control system
- Fluid power system
- Process control system
- Closed circuit TV system

Fig.1 shows the Fuel Handling Control System block diagram. The computer system is a distributed control system consisting of a mini-frame computer as the master and two microcomputers as slaves. The master computer does all the supervision and performs system scheduling and maintenance tasks. It passes codes to the micros for execution of process tasks and checking of permissives is also done by the micros. The commands for operation of field devices are given by the micros to the manual logic.

The electrical and electronic control system consists of the field devices feedback system, electric motor drive system and the safety and manual logic system. The commands from the micros ar checked in the safety logic system for safe operation before being sent to the fluid power system. The fluid power system operates the device in the correct direction, at the required speed and force, to position it accurately. The feedback device (say potentiometer) is read by the micro and appropriate commands initiated. The device is also connected to the computer independent display on the console. Operation and alarm messages are diplayed on the VDU and logged by the printer.

One of the important requirements of the fuel handling control system described earlier is that the operator should be able to take over the control of the machine should the situation warrant or during upset conditions (say after a seismic disturbance). Since it has been difficult to get computer systems that would be able to operate in the environment after an upset condition, the manual and safety logic has been built to meet these requirements. The options available to implement this logic were PLC, FPLA and SSI ICS. SSI ICS have been used to build the combinational logic as they afforded the simplest way of implementing the logic. They also met the other criteria of no mixing of logic, provision of bypassable intermediate signals, using a minimumm variety of cards and use of qualified components. This system has another major advantage that a fault in one portion of the system is not likely to affect the other portions of the system. However this has led to a large hardwired logic system.

The replacement of the manual and safety logic by a ruggedized micro computer system would make the system more compact. However replacement of the safety logic by software will require considerable research and development effort to achieve the required reliability. Also the validation of this software poses a real challenge.

However, it is proposed to use programmable logic assemblies in Kakrapar, to build the manual and safety logic. The system is not logic intensive and seggregation of devices will involve using a larger number of assemblies than absolutely necessary; yet on a review it has been found that distinct advantages will accrue by using these assemblies. These advantages are a) reduction in the overall cost b) reduction in the variety of PCBs and consequent reduction in engineering and manufacturing effort c) reduction in the amount of wiring.

(b) Operator-information system

The operator information system has been computerized in Narora. Most of the data logging and displays are provided by the computerized system. To reduce the strain on the operator a number of computer driven CRTs have been provided. Some of these provide the needed information in the form of bar charts, trends, or in tabular form, on operator´s demand. The critical parameters are displayed automatically. A few of th CRTs are used exclusively for the purpose of alarms.

Hard copies of log of alarms and history plots etc. can be obtained. This system provides all the information, with minimum continuous display, thus relieving the operator as he does not have to look at unwanted information. Also since a programmable system is available, it will be easier to change the priority of displays and include the desired parameters in a particular table whenever necessary. Under upset conditions the event can be analysed readily.

Some of the critical parameters, specially those relating to the safety of plant, are still provided on analog meters, recorders and window annunciators for alarm .

(c) Channel Temperature Monitoring System

In the earlier reactors, in one of the two redundant installations, computers were used. In NAPP both the installations have been provided with microprocessor based systems. The alarms for unsafe conditions are however generated as in earlier reactors.

(d) Reactor Regulating System

In the earlier reactor the regulating system was a hard wired analog system. The basic building blocks for the system were power demand unit, amplifier/processing system and median deriving circuits. The reactor power demand is entered in the power demand system. The amplifier/processing system generates an error after comparing the demand with the actual power. The median circuit selects the median of the three channels to control the reactivity. The old system is shown in figure-2. The new microprocessor based system provides all the functions through software. The advantage of the system are apparent. The system processing function can be changed easily through software. It provides facility for in-situ system parameter adjustment and change of control strategy to provide the best dynamic and static performance. Figure-3 shows the system block diagram. Figure 4 shows the system flow chart.

There are three identical reactor regulating channels, each having a provision for median derivation. The system has been designed such that failure of any one channel, automatically transfers the control to the remaining two. For this change over relay circuits have been used.

(e) Other Areas in C & I Systems using Microprocessors

Microprocessors have also been used for local area radiation monitoring system, heavy water leak detection system and failed fuel detection system.

The field of turbogenerator controls provides scope for use of programmable electronics for future reactors. Upto Narora most of the safety and control logics have been hardwired.

FUTURE STRATEGY

As seen, a trade off between the analog, computer based and other programmable electronic systems is being done continuously. However, in the case of protective system and certain critical logic systems relays are still in use. The basic problem, specially for protective system, is about qualifying the software. It appears that relay logic will continue to be used for quite some time in the protective system. But for other logic systems programmable devices/logic controllers will gradually replace the relay logic circuits. Also extensive use of programmable systems will be made in the field of data logging and annunciation system.

Use of programmable system is gradually increasing. However, presently it is largely confined to local systems configuration. In future reactors while still maintaining local sub-systems, a supervisory main frame computer to communicate with the sub-systems will be introduced.

CONCLUSION

Programmable electronics will gradually replace the existing hardwired system in the Indian PHWRs. However, in certain areas like protective system a more cautious approach towards this system has to be taken.

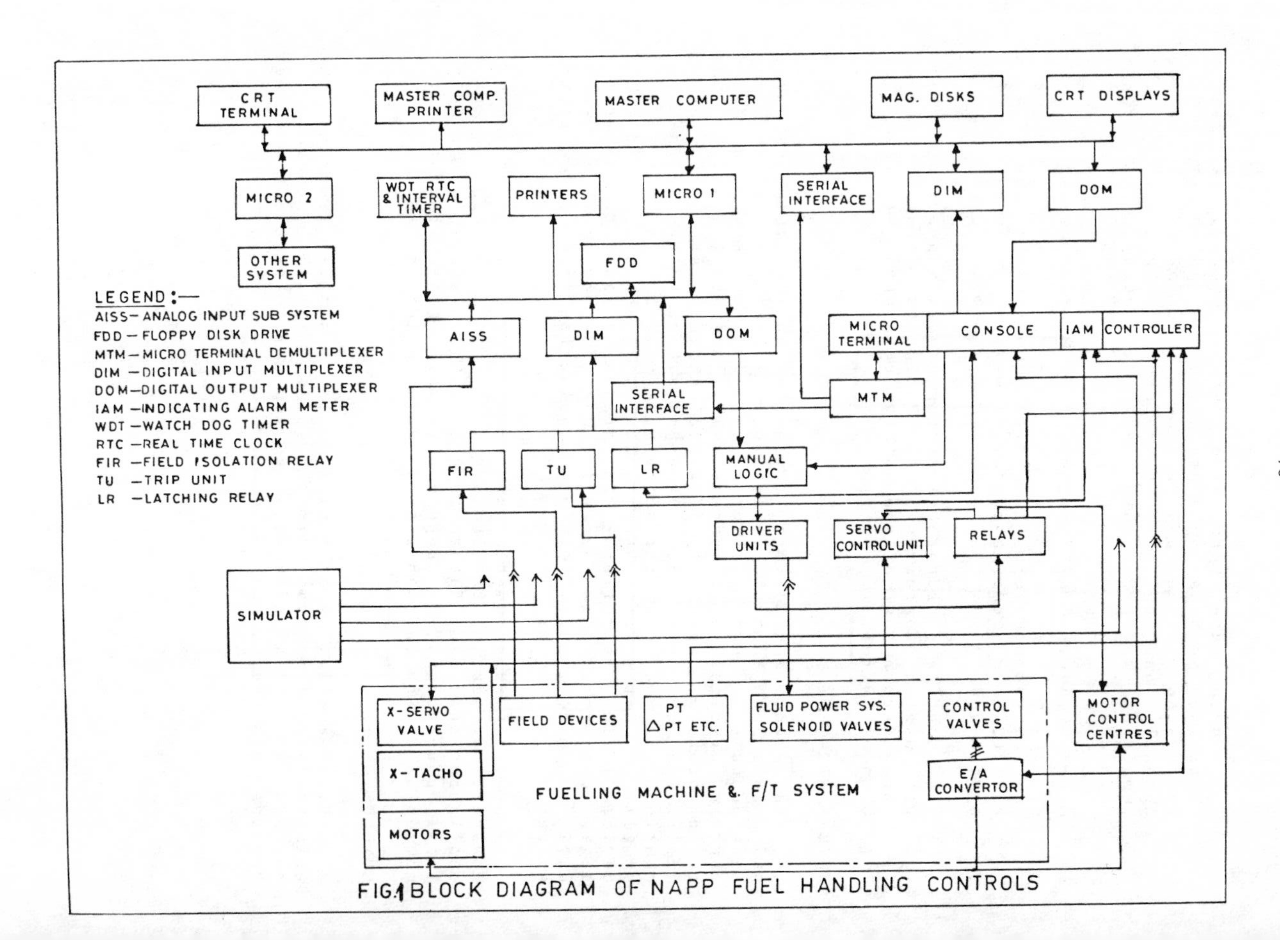

FIG.1 BLOCK DIAGRAM OF NAPP FUEL HANDLING CONTROLS

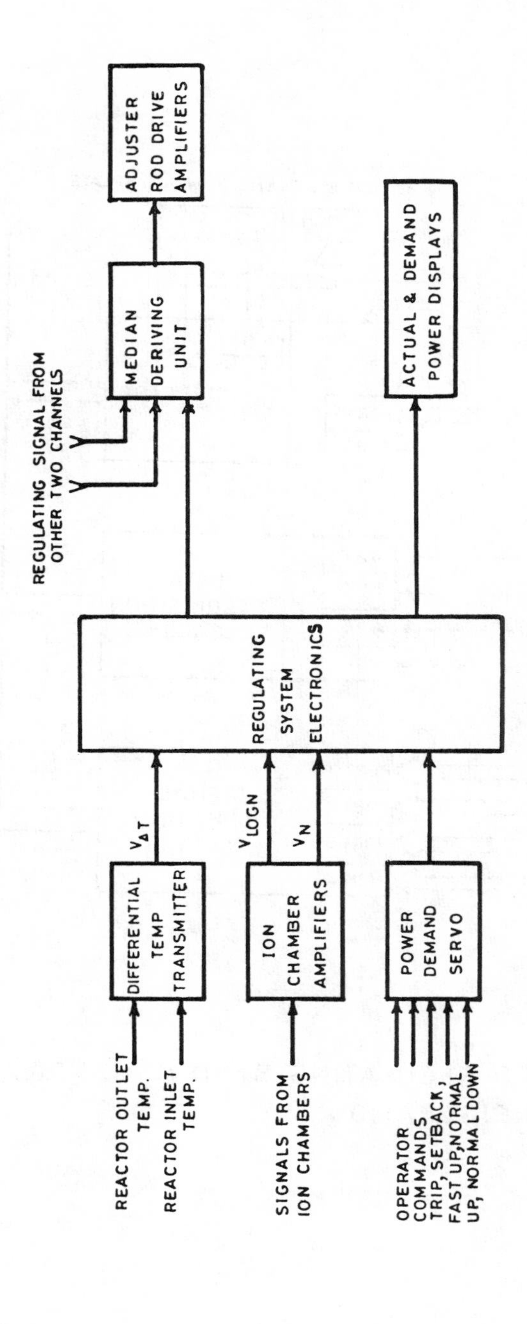

FIG.2 REACTOR REGULATING SYSTEM BLOCK DIAGRAM

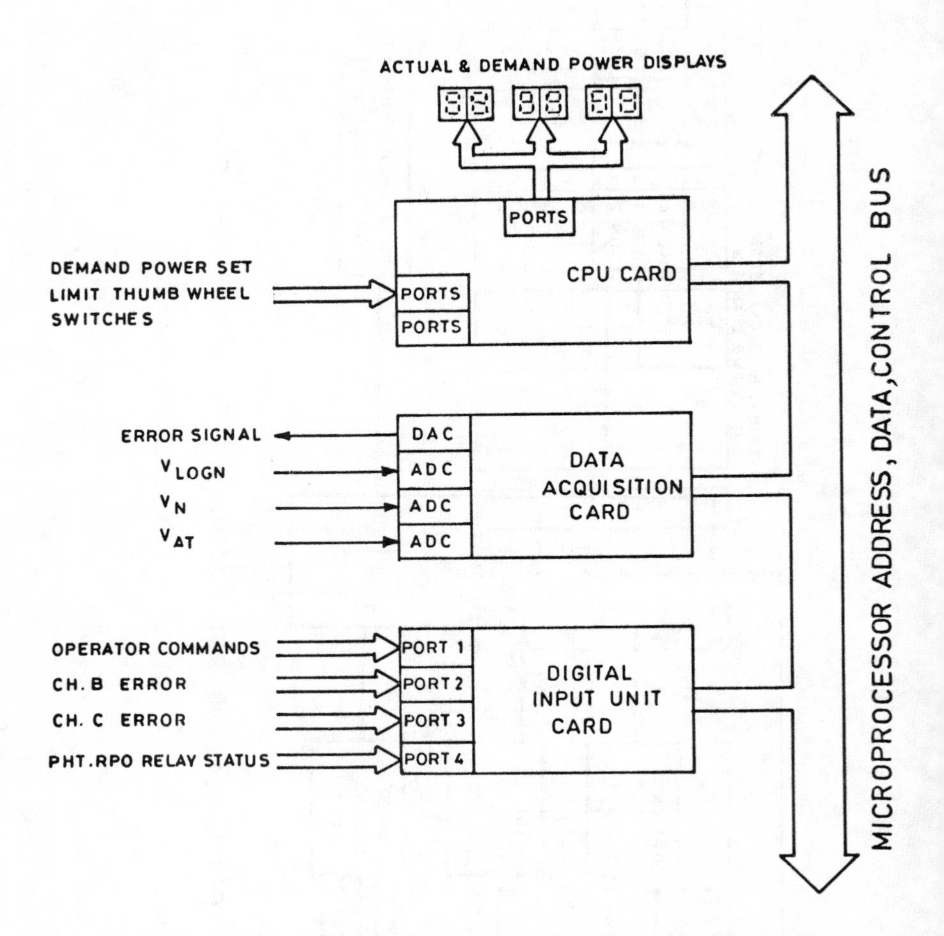

FIG.3 NAPP REGULATING SYSTEM HARDWARE CONFIGURATION

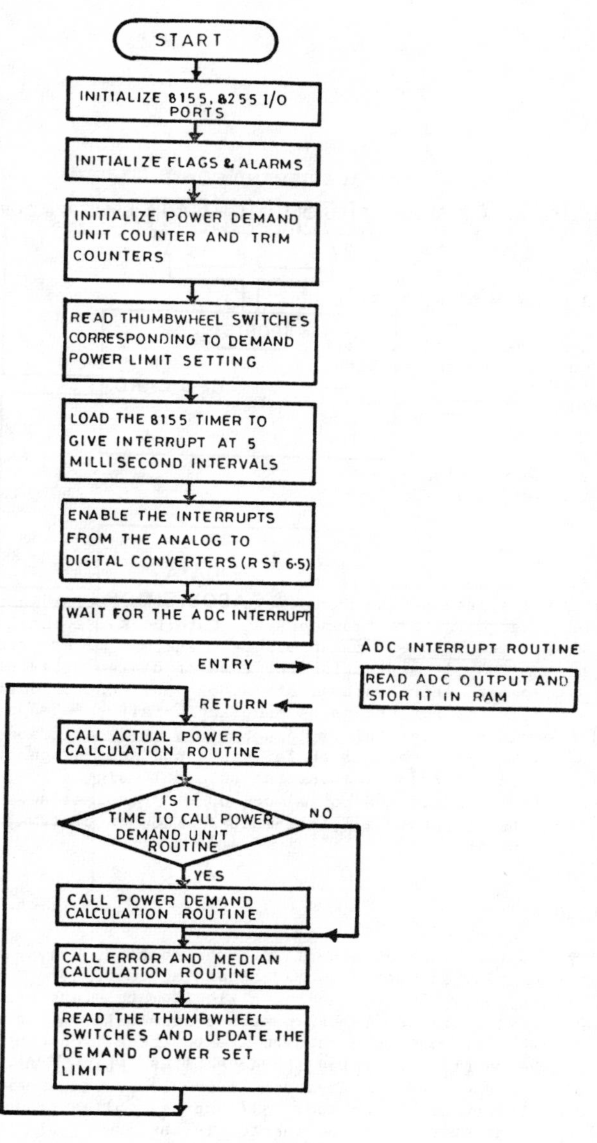

FIG. 4 FLOW CHART OF SYSTEM MONITOR (REGULATING SYSTEM)

FAILSAFE OPERATION OF A PES IN A LIQUID METAL FAST BREEDER REACTOR REFUELLING SYSTEM

LAWRENCE L A J, PULLEN D and SMITH I C

United Kingdom Atomic Energy Authority,
Winfrith, Dorchester, Dorset, DT2 8DH, UK

ORME S

National Nuclear Corporation,
Risley, Cheshire, WA3 6BZ, UK

ABSTRACT

The refuelling system of the UK Commercial Demonstration Reactor (CDFR) includes a large number of sensors and interlocks, a considerable proportion of which are vital to plant safety. The use of Programmable Electronic Systems (PES) is attractive for ease of design, flexibility and reliability. The paper describes some of the onerous control duties required and stresses the advantages of adopting a fail-safe approach. Conforming with earlier successful implementations, a dynamic mode of operation is advocated from which both fail-safe and self-diagnostic properties derive. A laboratory demonstration model using a commercially-available PES to control a mock-up of a typical automated operation is described in which a Triple-modular Redundancy (TMR) configuration is employed.

1 INTRODUCTION

The heat-producing heart of a nuclear power station, known as the 'core', consists of many elements; some containing fuel, some containing absorbing materials, etc. In the case of a liquid metal fast breeder reactor (LMFBR) these elements are known as sub-assemblies. The three types of sub-assembly of most concern here are the heat producing fuel sub-assemblies, the neutron absorbing sub-assemblies (which are used for reactor control) and the breeder sub-assemblies which produce new fuel due to the effects of the nuclear reaction. All sub-assemblies produce a certain level of decay heat after the reactor is shut-down, which is of vital concern during the refuelling operation.

In order to maintain the nuclear reaction within the core it is necessary to refuel the reactor. This consists of replacing burnt-up sub-assemblies with corresponding new sub-assemblies. In the case of the CDFR (the Commercial Demonstration Fast Reactor) six months is used as the period between refuelling shutdowns. During a refuelling shutdown up to one-third of the 600 sub-assemblies within the core are replaced.

The problems associated with carrying out the refuelling operation require a high integrity control system for their solution. The two key criteria for such a control system are (a) safety and (b) system availability. The safety aspects cover areas such as prevention of radioactive release and plant damage whereas system availability includes reduction of downtime and hence reduction in lost revenue due to the refuelling shutdown.

The aim of this paper is to outline a method of using 'off the shelf' programmable electronic systems (PES) in a failsafe redundant configuration with which it is hoped to achieve the safety and availability requirements for a refuelling control system.

2 CDFR REFUELLING SYSTEM

The CDFR refuelling system envisaged is as follows (1):-

(i) IN-VESSEL HANDLING

This covers all plant within the reactor primary vessel, or mounted on the rotating shield, except the bucket erection system.

Two identical charge machines (direct lift) are carried on the inner shield of the two-rotating shield system located above the reactor core (Fig 1). One is designed to handle irradiated core components, the other new; both, however, are designed to remain in the reactor environment during power operation, although component transfers only take place at shutdown.

In order to replace a core component, the empty charge machine lifts out the component until it is clear of the core and within the charge machine chute. The shields are then rotated until the charge machine with the new component is above the empty core position; the new component is then deposited in that position. The shields are again rotated until the charge machine with the irradiated component is over the bucket station. The irradiated component is then deposited into a bucket, the component transferred out of the reactor vessel and a new component transferred into the vessel. The new component is then accepted by the charge machine and the cycle repeated.

(ii) TRANSFER SYSTEM

This covers all plant associated with the ex-vessel transfer of core components between the bucket erection system in the reactor primary vessel and a similar bucket erection system in the core components store situated inside the core components preparation block and outside the reactor secondary containment building.

Core components are transferred from the bucket station in the reactor to the bucket station in the core components store (CCS) (Fig 1), which is located outside the reactor containment building. The transfer system consists of two inclined ramps; one entering the reactor and one entering the core components store. The lower ends of the ramps, both in the reactor vessel and the store, have a fixed cam for moving the fuel bucket into a vertical position for the insertion and removal of a sub-assembly. The upper ends of each ramp are connected into a compact fuel transfer cell which contains a rotor for translation of the fuel buckets from the reactor to the CCS ramp and vice versa. The buckets are moved along the ramps by two winches, which are permanently connected to the buckets by chains, and hoist the sodium filled bucket containing the sub-assembly out of the sodium pool.

The thermal capacity of the bucket is sufficient to maintain fuel cladding at less than the on-power operating temperature during the time taken for normal transfer between the reactor vessel and the CCS. To cater for the risk that a hold up occurs while the fuel is out of the sodium pools, the bucket will be finished with a profile which enables the heat to be transferred to the ramp tubes by radiation. A steady state heat removal capacity of about 35 KW can be achieved within a temperature limit which will not jeopardise the fuel pin integrity. By switching on a gas cooling system the heat removal capacity can be enhanced.

To avoid delaying refuelling for the fuel decay heat to fall to this limit, sub-assemblies will be changed in order of increasing initial decay heat power. Automated systems will be used to control fuel handling operations with built-in safeguards to prevent faults. These safeguards include control over sequencing of fuel handling, direct identification by the ultrasonic identifier reading a code built into the top of each sub assembly, measurement of gamma and neutron emission from the fuel and a means of locking out the refuelling equipment from specific core locations until sufficient time has elapsed for heat output to decay to a safe level.

(iii) CORE COMPONENT STORE

The CCS is the storage facility to accommodate new and irradiated core components, including fuel, the latter until the decay heat has reduced to a level acceptable for further handling. The CCS, which is sized to accommodate a complete fuel charge, is contained in a steel vessel filled with sodium (200-250°C) surrounded by a leak jacket which will maintain a sodium level above the fuel in the unlikely event of leakage from the storage vessel.

Two offset arm charge machines are used within the store for transfer of fuel between the bucket and storage positions. Fresh fuel is loaded into and irradiated fuel is removed from the store by means of a flask which transfers the fuel between the store and cave facilities thus ensuring complete segregation of the fuel store from the caves.

3 SYSTEM RELIABILITY AND AVAILABILITY CONSIDERATIONS

3.1 FUEL HANDLING SYSTEM FAULTS

The faults relating to the refuelling system fall into the following categories:-

i In-reactor fuel handling faults

a. selection and transfer of an overrated fuel S/A
b. inadvertent shield rotation
c. dropped core component

ii Fuel handling faults in the fuel transfer system

a. inadequate cooling for fuel meltdown of a 35 kW fuel S/A during transfer
b. inadequate cooling for an overated fuel S/A during transfer
c. dropped loads

iii Fuel handling faults in the core components store

a. inadvertent rotation of charge machine or store matrix during refuelling
b. dropped core component.

Several of these faults will influence the control system from the point of view of safety or plant downtime. The fuel route control system interlocks are laid down by the CEGB in Reference 2, which also quantifies the recommended reliability for different classes of interlock. A discussion on the achievement of the recommended reliabilities is beyond the scope of this paper, which will confine itself to the role of fail-safe, self-diagnostic PES design, and their general influence on reliability considerations.

The arguments for a high availability system are based on the loss of revenue due to reactor loss of generation. At present the figure for replacement energy lost is about £20 per MW-hr [≈ £600,000 per day generating at 1300 MW].

3.2 CONTROL SYSTEM REQUIREMENTS

A control system is required for the CDFR refuelling route which has two important qualities: firstly, a high availability in order to minimize control system downtime, and secondly a low probability of failing on demand. Whereas the first is linked with minimizing the cost of a refuelling shutdown, the second is linked with avoiding the initiation of a potential hazard. Also the question of minimizing hardware and design costs consistent with satisfying the requirements is an important consideration.

It is clear that many "off the shelf" commercially-available programmable logic controllers have the capability to carry out the desired sequencing and interlock duties, most having extensive self-testing/diagnostic features included. However, these were felt to be insufficient for the required duties and the programme of work described below was initiated with the aim of achieving a high degree of fail-safe operation as a design goal.

3.3 FAIL-SAFE, SELF DIAGNOSTIC PROPERTIES

In high-integrity control and protection systems, some failure-survival capability is derived from the use of redundancy ie the replication of the system hardware by constructing a number of active identical subsystems and combining and voting on the subsystem outputs. In this way, the system can survive either a failure-to-safety or a failure-to-danger in one or more subsystem and continue to operate as designed. If in a voted Triple Modular Redundancy (TMR) protection system, for example, a logic 1 is defined as representing the normal, healthy state, and a logic 0 as the abnormal, unhealthy state, then a hardware fault leading to a "stuck-at-0" would translate as "safe" fault whereas a fault leading to a stuck-at-1 would translate as a "dangerous" fault. With majority-voting, the system could survive a single "stuck-at" fault, either avoiding a spurious trip, or avoiding a loss of protective capability without any human intervention. However, there is a limit to the number of simultaneous failures which can be tolerated. These may occur either as common-mode failures, or as an accumulation of random failures, either of which may be of a revealed or an unrevealed nature.

A consideration of these factors leads to the conclusion that fail-safe design is good practice in systems requiring high integrity ie designs in which not only do (a) the large majority of failures (ideally all) lead to a vote for protective action, but (b) also are automatically made apparent by design.

The properties (a) and (b) have the following advantages:

i Common-mode hardware failures are more likely to lead to a safe consequence than in non-failsafe designs.

ii The frequency of independent dangerous failures is likely to be reduced.

iii If failure-to-safety is combined with automatic indication of the failed-safe state, prompt repair action may be instigated, thereby minimising the outage risk. If the hardware is designed on a modular basis, with modules incorporating failure-warning indicators, the diagnostic time is minimised.

A big step forward in fail-safe design can be made by replacing the simple "DC" concept of logic 1 = normal, logic 0 = abnormal, by the concept of alternating between 1 and 0 ("dynamic") as the normal state and "static" as the abnormal state. In this way stuck-at faults lead to a change of state from dynamic to static and subsystems can be designed so that the interpretation is simple and unambiguous. Examples of protection systems employing dynamic operation are: LADDIC (reference 3), semiconductors employing pulsed-coded logic (reference 4) and self-testing microprocessor-based safety systems (references 5, 6).

Considering random, independent hardware failures the two essential parameters in an s-out-of-n voted-redundant system are

(a) The fractional dead-time (mean probability of failure to respond to a demand for protection) = μ_D.

(b) The spurious trip rate = Λ.

The term "trip" is used in this paper, since it is in such common usage. However, in the context of fuel handling systems it may also be interpreted as an automatic initiation of an inhibiting action (prevention of a given manoeuvre).

Parameter (a) is derived in reference 7 as

$$\text{Fractional dead-time, } \mu_D = {}^{n}C_r \frac{\theta^r \tau^r}{r+1} \quad \ldots (1)$$

where n = number of redundant subsystems
r = number of dangerously-failed systems to cause dangerous system failure (= $n-s+1$)*
θ = subsystem fail-danger rate
τ = test interval to check for fail-danger faults.

Parameter (b) can be shown to be

$$\text{Spurious trip rate, } \Lambda = {}^{(n-1)}C_{(s-1)}\, n\lambda^s (\tau_{rep})^{s-1} \quad \ldots (2)$$

where s = number of failed-safe subsystems to cause safe system failure*
λ = subsystem fail-safe rate
τ_{rep} = mean time to repair fail-safe faults.

For a 2-out-of-3 system, equations (1) and (2) give

$$\mu_D = \theta^2 \tau^2 \quad \ldots (3)$$

$$\Lambda = 6\lambda^2 \tau_{rep} \quad \ldots (4)$$

For given values of θ and λ, the fractional dead-time and spurious trip rate are controlled through the test interval τ, and the mean repair time τ_{rep} respectively.

Clearly, there are constraints on both these parameters, since a test for dangerous failures may only be possible during a maintenance shut-down, and the time for on-line repair of safe failures may be limited by operational considerations, staff availability etc.

However equations (3) and (4) cannot be regarded as anything but helpful guides, since in cases where highly reliable systems are sought, common-mode failures will dominate: errors in maintenance, errors in design, environmental conditions etc. (Reference 8); software design errors (Reference 9).

---

* For the sake of simplicity the analyses are confined to safe, revealed faults and dangerous, unrevealed faults.

4 IMPLEMENTATION

In order to demonstrate dynamic operation in TMR control systems, a standard off-the-shelf GEC GEM 80 PES system together with appropriate ancilliary equipment was acquired as indicated in the system diagram of Figure 2, and programmed to operate in conjunction with a demonstration model designed to emulate a typical manoeuvre in a fuel-handling system.

For the demonstration model, the sub program for the Transfer System was selected, since this is the most critical system from the safety view point. This deals with the raising of a bucket of spent fuel from the reactor, and of a bucket of new fuel from a fuel store, to a transfer cell where the two buckets change places. They are then lowered, the new fuel to the reactor and the spent fuel to the store.

The two hoists and the transfer cell were represented by motor-driven mechanisms, with triple-redundant microswitches sensing

1. Reactor hoist down
2. Reactor hoist up
3. Store hoist down
4. Store hoise up
5. Transfer cell operated.

and providing appropriate enabling signals for the automated operations.

For the purpose of demonstration, TMR operation was simulated by running the same program sequentially in response to appropriate input switch commands from the Simulation Panel and the microswitches. The respective outputs were obtained as described below. For convenience, the "redundant" outputs from the PES were voted, using conventional hard-wired logic.

To achieve dynamic operation in the PES, for the reasons given in 3.3 above each of the "three" controllers were programmed to alternate between the "on" and "off" state with a fixed phase relationship to one another. This was implemented by invoking the sequencer facility of the GEM 80. This is an instruction sequence by which an internal register may be nominated as a single bit shift register. For the first pass of a program using this function, bit "0" of the selected register is set. For succeeding passes the bit is moved to progressively higher order positions in the register, returning to bit "0" after 16 passes or after reaching a position programmed as the reset bit.

The program is such that 6 output phases of equal duration are produced. These are arranged as follows:

PHASE	0	1	2	3	4	5
Controller A	On	On	On	Off	Off	Off
Controller B	Off	On	On	On	Off	Off
Controller C	Off	Off	On	On	On	Off

If all 3 controllers are functioning correctly the output of the 2-out-of-3 gate is:-

PHASE	0	1	2	3	4	5
Output	Off	On	On	On	Off	Off

The outputs for the various fault conditions are best illustrated by a timing diagram as shown in Figure 3. The failure of one controller to the "on" state and of another to the "off" state would result in the output of the remaining controller appearing unmodified at the output of the 2-out-3 gate. This difficulty may, however, be removed by employing transformers to couple the controller outputs to the 2-out-of-3 gate inputs.

6 passes of the program being required to produce each cycle of output, the minimum period using the GEM 80 130C type of controller is 80 mS. However, a sixfold speed increase would accrue if for example the more modern GEM 80 series controllers were used.

It will be noted in Figure 3 that either more than one controller failure, or more than one change of state in the input signals results in a static condition at the output of the voters. In this application therefore this corresponds to the "unhealthy" state in which actuation of the mechanisms and therefore movement of fuel is inhibited, representing a "fail-safe" capability for those parts of the controllers concerned with executing the programs, the outputs and the voting logic, but of course excluding the input modules and the microswitches. Also, logical interpretation of the waveforms associated with single failures enables failure diagnostics to be performed, ie the faulty subsystem may be identified.

5 CONCLUSIONS

5.1 In addition to providing a defence against common-mode hardware failures, failsafe design, if combined with self diagnostic capability can improve system reliability by

(a) reducing the frequency of dangerous failures and therefore for a given voting system and risk time, the fractional dead-time.

(b) reducing the spurious trip rate through the medium of on-line repair of failed-safe units.

5.2 Using an off-the-shelf GEC GEM 80 Programmable Controller, together with various necessary peripherals and an electromechanical model of a representative part of a CDFR fuel handling system, an exercise was carried out in which the PES was programmed to execute various defined manoeuvres, but in such a way that dynamic operation in the PES was instituted. To save cost, TMR redundancy was simulated by successive operations of the same program. The operating mode was such that given that the fuel handling "components" were in the correct state and the PES in a given "subsystem" healthy, a dynamic drive signal was output, and voted on. Failure of the subsystem, or of the components to be in the correct state, caused the dynamic output to become static. Thus a very large proportion of hardware failures in the PES would cause the change of state from dynamic to static ie the performance was "fail-safe".

5.3 The phase relationships of the "subsystems" were coded in such a way that the voted waveforms provided diagnostic information to identify the PES giving a static output. For convenience, hardwired voting was used. In this way the voted PC outputs were interpreted as

dynamic ≡ drive
static ≡ drive vetoed

The frequency of the square wave corresponding to 'drive' was however rather low (12.5 Hz) for practial purposes. This could be resolved by using other, faster-operating PES.

REFERENCES

1. J A G Holmes. 'Developments in UK Fast Reactor Design'. BNES Journal, February 1981.

2. CEGB Standard 500117 'Principles of Nuclear Fuel Route Interlock and Control System'. GDCD Standard 194. April 1980.

3. A H Weaving and J Sherlock. "Magnetic Logic Applied to Reactor Safety Circuits", J BNES, January 1963, 74.

4. A B Keats. "A Trial Pulse-coded Logic Automatic Shutdown System", IAEA Symposium on Nuclear Power Plant Control and Instrumentation, Munich 1982, Paper No IAEA-SM-265/29.

5. A B Keats. "Failsafe Design Criteria for Computer-based Reactor Protection Systems", Nuclear Energy 1980, 19, Dec, 6, pp423-428.

6. S Orme, N J Evans and B O Wey. "A Fail-safe Microprocessor-based Protection System utilising Low-level Multiplexed Sensor Signals". J Phys. E: Sci. Instrum., Vol 18, 1985.

7. A E Green and A J Bourne. "Reliability Technology", Wiley Interscience 1972.

8. J Bourne et al. "Defences Against Common-mode Failures in Redundancy Systems - a Guide for Management, Designers and Operators", SRD - R196.

9. P Bishop et al. "Project on Diverse Software - An Experiment in Software Reliability". Proc. 4th IFAC Workshop, Como, Italy, 1-3 October 1985. Pergamon Press.

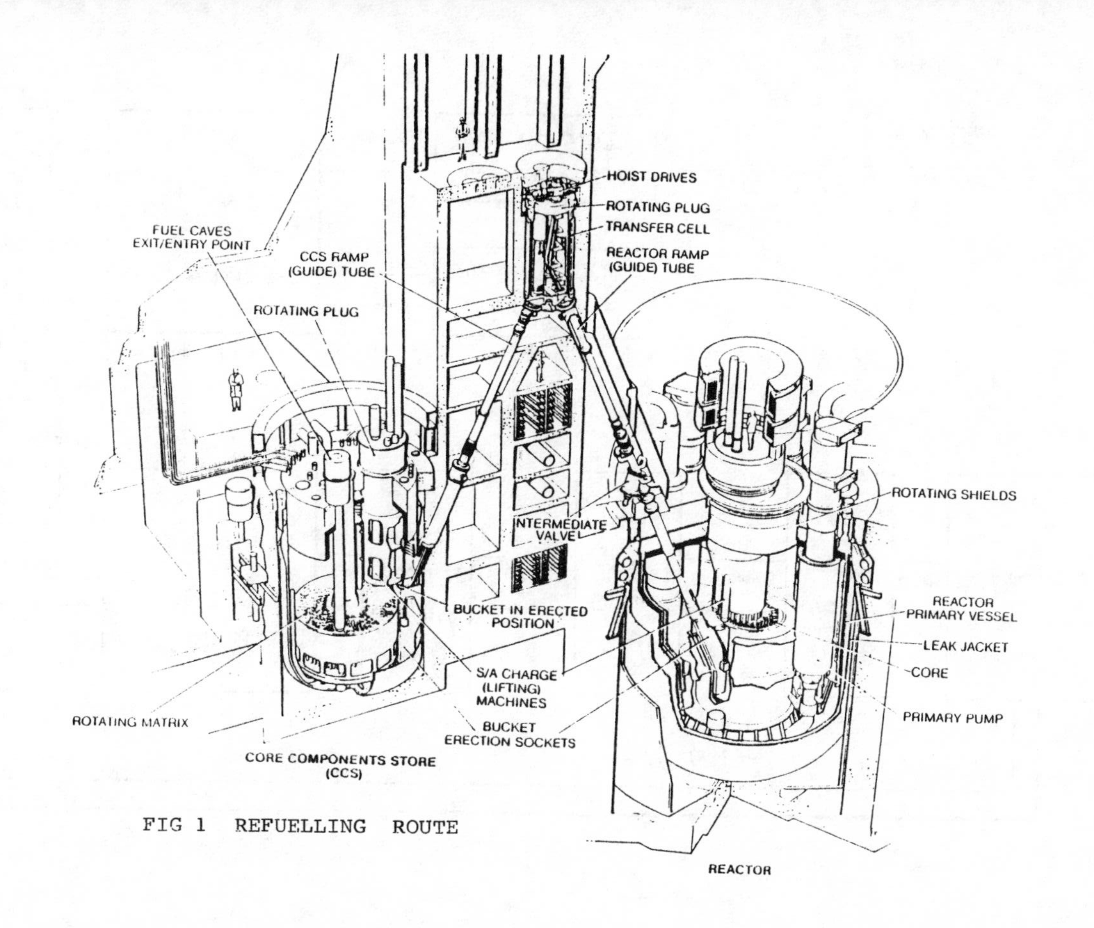

FIG 1 REFUELLING ROUTE

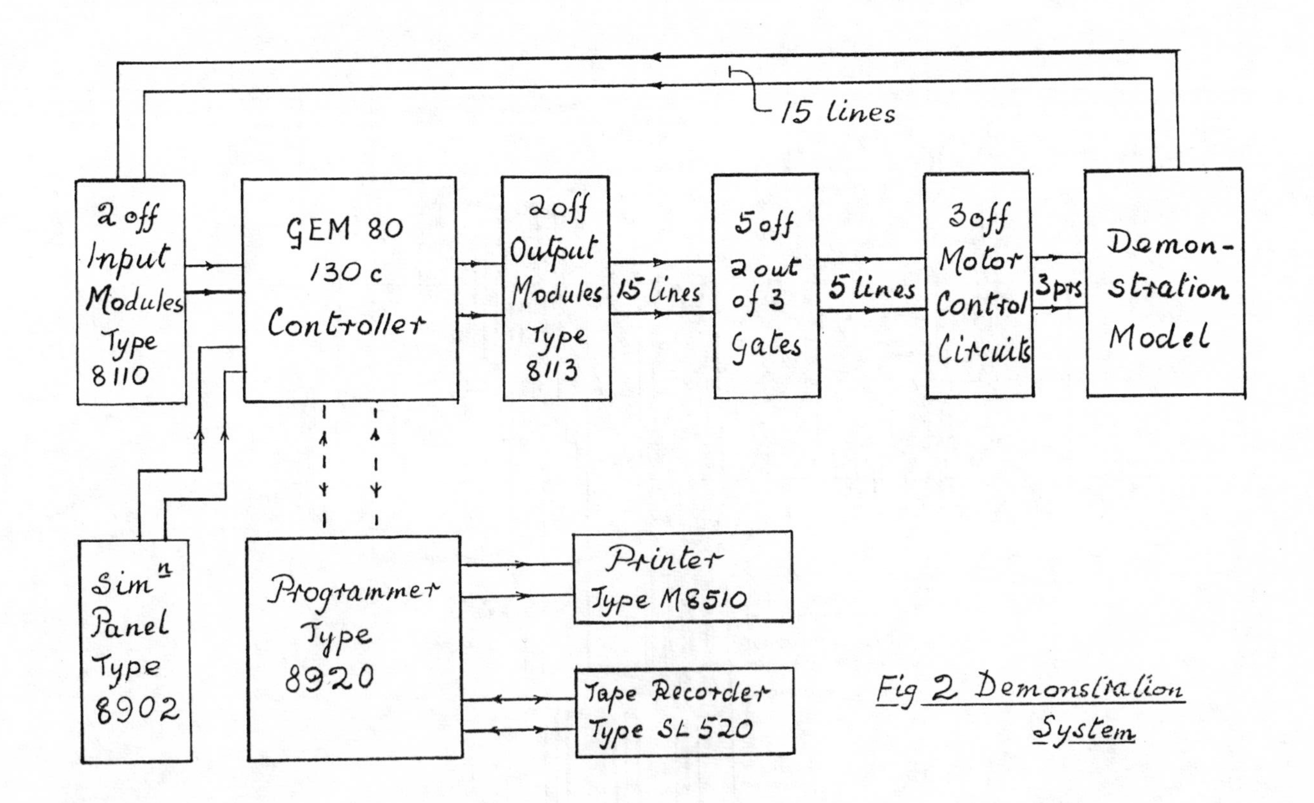

Fig 2 Demonstration System

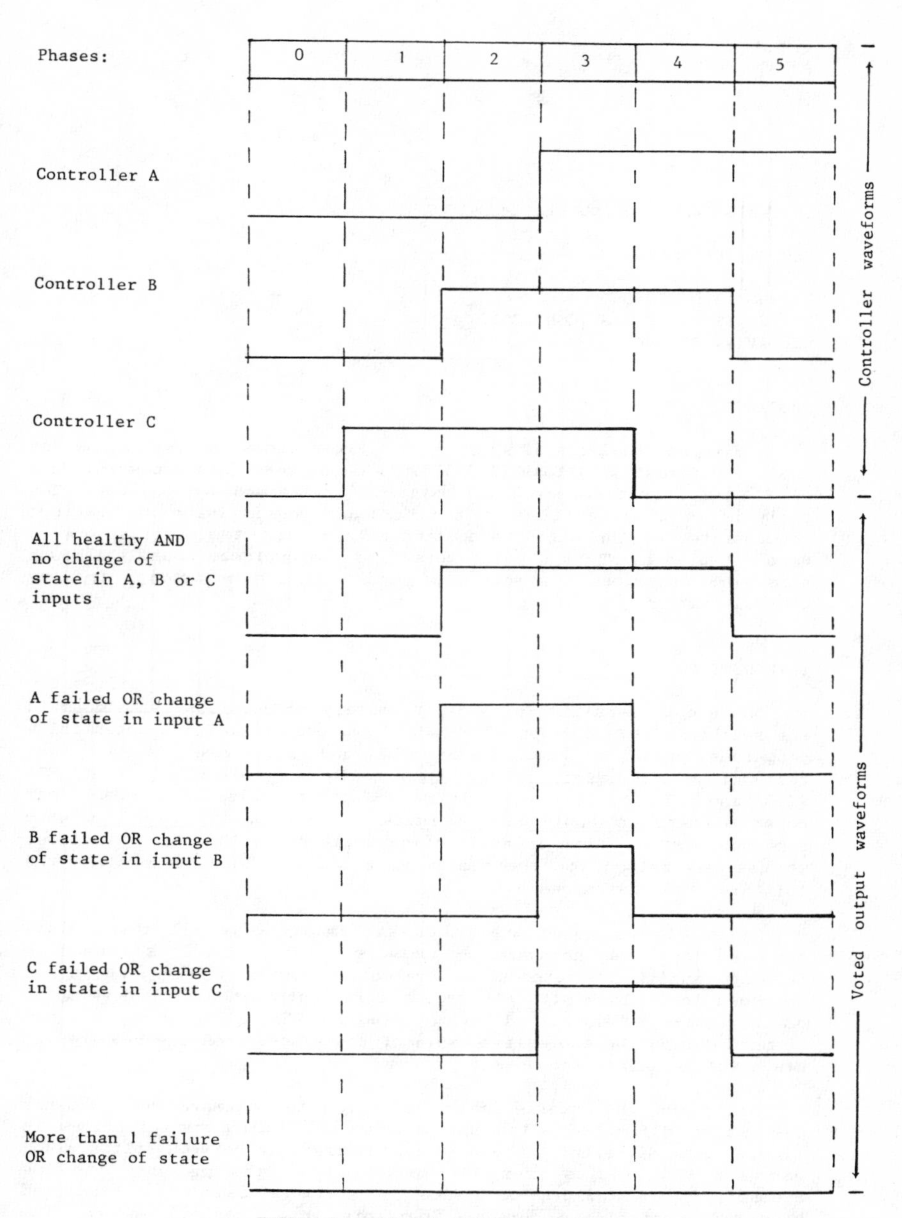

FIGURE 3: TWO-OUT-OF-THREE SYSTEM: CODING DIAGRAM

SOFTWARE SAFETY USING FTA TECHNIQUE

REUVEN GREENBERG

LICENSING DIVISION
ISAREL ATOMIC ENERGY COMMISSION
TEL-AVIV, ISRAEL

ABSTRACT

Fault Tree Analysis (FTA) is a well known and effective method for analyzing hardware systems {1-3}. This paper describes a possible use of FTA in a software embedded system for temperature control. The analysis is first applied to the hardware and software systems but concentrates on the software section afterwards. two critical events were detected by FTA and steps were taken to overcome them. Although this work describes a specific situation it can be applied to many control systems.

INTRODUCTION

In recent years it has become necessary to develope techniques to ensure the safety of computer embedded systems controlling potentially dangerous processes. Some works published last years showed that reliability and safety improvements could be achieved by using FTA {4,5} and FMEA {6} (Failure Mode and Effect Analysis) techniques. This paper attemts to apply and evaluate the FTA method in a software embedded system. Such an application will enable the safety engineer to use one method for the system as a whole without separating the software from the hardware.

It should be noted that the FTA cannot solve all the prblems involved in either hardware or software safety. There is a need to develope additional methods for evaluating system safety. It is not intended to replace with FTA proven efficient methods for developing software safety {7}, but rather to integrate FTA within them. FTA can be used during the whole life-cycle of the system from the requirement stage to the maintenance stage.

Since the characteristics of failures in software and hardware systems do differ {8}, the approach to overcoming such failures is likely to be different. The principal difference between software and hardware FTA results from the fact that at the present time the probability of failure occurence in software cannot be predicted within acceptable accurecy. Therefore one cannot ignore the possibility of software failure by claiming that it is unlikely to occur.

FTA FOR IMPROVING SAFETY

Safety failure can be defined as the inability of a system to perform a required function which leads to serious or dangerous consequences {4}. Since the interpretation of "serious consequences" is left to the designer, user, etc. he may define a failure to be critical even though it is not considered as such. How can a designer overcome a critical failure? There is no unique answer to this but rather a variety of solutions, depending on specific event and requirement. Most safety criteria require a system to be just fail-safe; However in some cases self-recovery or prevention is required. Hence the final solution is dependent upon the requirements of the individual system.

The FTA consists of five {3-5} principal stages:

1) Determine the critical failure(s) (Top-event).
2) Assume that the failure already occured.
3) Locate the event which may have caused the top-event to occur.
4) Follow the chain of events untill the primary ones are detected.
5) Attempt to overcome the primary events (In most software failure this is the only solution) or prove that the probability of their accurring is low enough to meet the safety criteria.

TEMPERATURE CONTROL SYSTEM

Let us assume a simple negative feedback system for controlling temperature in a cold environment. The control unit samples the temperature periodically* and monitors the output power unit which is responsible for the heating. as the system is in a cold environment it loses heat to its surroundings. At equilibrium the amount of heat supplied to the system equals the amount lost.

The power (W) supplied to a system at teperature-T which ought to be at temperature T_o is given in equation (1):

$$W = W(T_o) + c(T_o - T) \qquad (1)$$

Where c is a constant known for a given system and $W(T_o)$ is the power needed to maintain the equilibrium at temperature T_o. The first right term of eq. (1) is always positive and the second one is positive during the heating process and negative during the cooling process. The values of $W(T_o)$ as a function of T_o have been previously fixed by experiance or calculation and are stored in a "look-up-table". Intermediate values not in the table are interpolated.

---

* The periodicity of sampling must be very reliable to meet both safety and functional requirements. This is ensured by standard methods (watch-dog devices, etc.). Therefore it is assumed that if such failures exist their source is identified as being in the hardware part of the system.

The system operates in two modes, automatic or manual:

Automatic- The system scans a temperature range automatically, stabilising the temperature at pre-selected points over a certain time period.

Manual- An operator sets-up for T_o and the system stabilizes on it.

SAFETY ANALYSIS

Relevant symbols used in this paper {3} are listed in Fig 1. Since software systems treats events which are "impossible to occur" or "occur always", this analysis will consider them by symbolizing the appropriate channel of the fault tree with "0" or "1" respectively.

In most cases thermal systems have two safety limitations:

1) They must not exceed a known temperature range from T_l to T_h.
2) They are sensitive to drastic changes in temperature so the rate of change in temperature $|dT/dt|$ must not exceed a known value (D_r).

These limitations are considered to be the most critical failures (top-events) and must be prevented. It must be noted that they are not the only ones to exist.

The upper or lower temperature bounds could have been exceeded [fig. 2(a)] if the supplied power was more than W_h or less than W_l. W_h or W_l is the power needed to stabilise the system at T_h or T_l respectively.

The failure channels described in Fig. 2(b) are caused by a rapid change in temperature. A rapid heating could be obtained when $W > W_f(T)$, where $W_f(T)$ is the power which makes the system at temperature T to be heated at a rate D_r or more. On the other hand, supplied power less than $W_d(T)$ [$W < W_d(T)$] would not prevent the system at temperature T from too rapidly cooling due to great heat loss to the surroundings.

Both Fig. 2(a) and Fig. 2(b) indicate that the failures could be attributed to either hardware or software faults. Hardware failures are out of the scope of this paper and will not be dealt with further. Since a software fault could cause the top events, the process responsible for obtaining W will be subjected to a thorough developement within the next two paragraphs

TEMPERATURE OUT OF RANGE

Figures 3(a) and 3(b) are a continuation of Fig. 2(a). The top event in Fig. 3(a) describes the sum of two values that should be greater than W_h [$c(T_o-T)>0$, heating process]. The above statement is true if either value is greater than W_h (left branch) or if both values are smaller than W_h but their sum is greater (right branch).

As the system stabilises T approaches T_o and $c(T_o-T)$ vanishes (more precisely becomes very small). Therefore the event $c(T_o-T) < W$

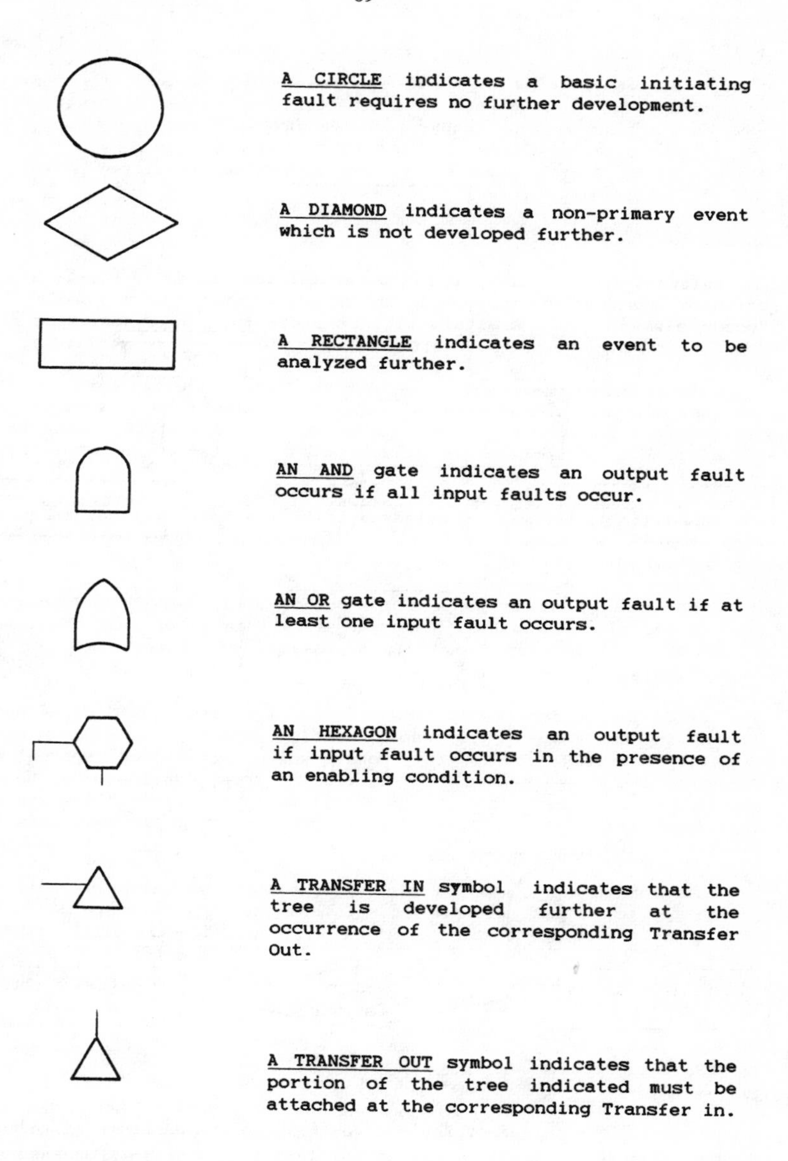

Fig. 1: Relevant fault tree symbols from Ref. {3}.

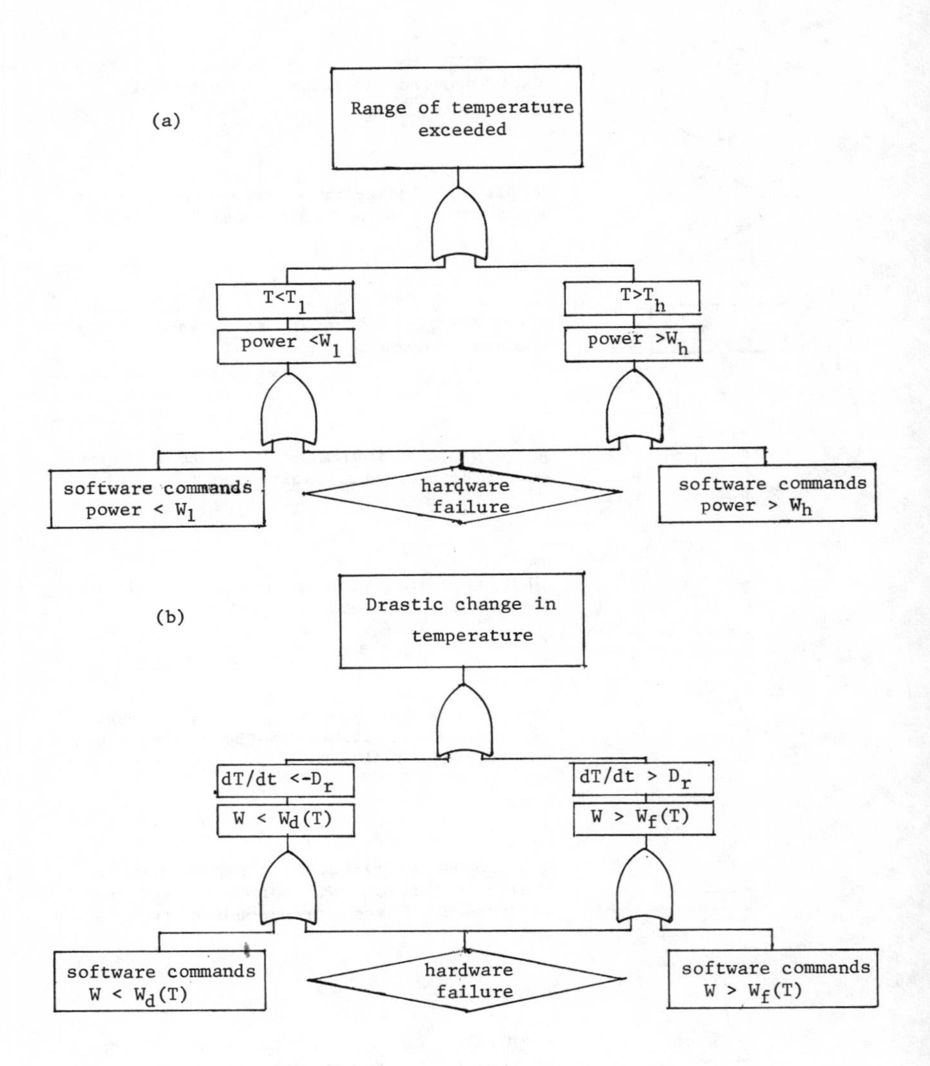

Fig. 2 - Safety faults: (a) range of temperature exceeded
(b) drastic change in temperature

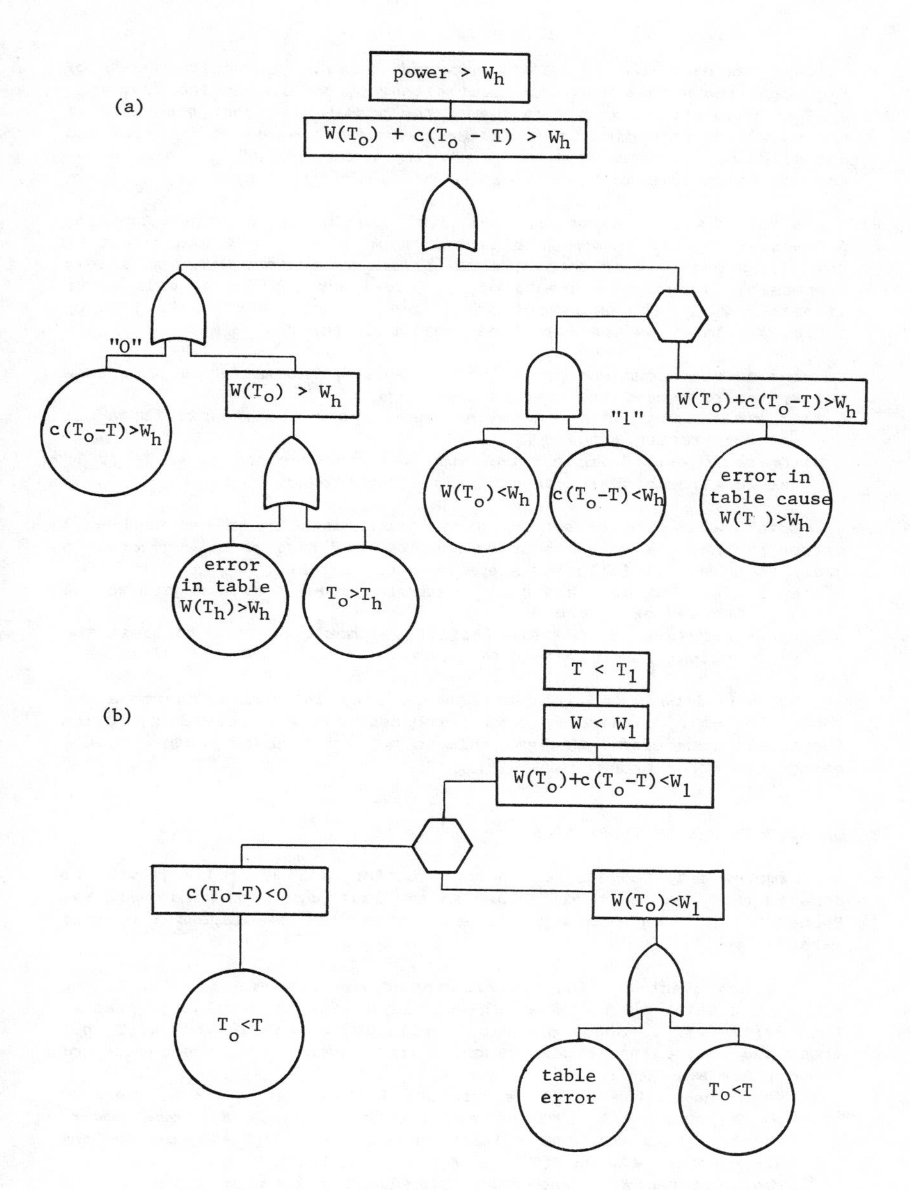

Fig. 3: (a) Temperature higher than T_h
(b) Temperature lower than T_1

"always occurs" and $c(T_o-T)>W$ does not occur. The left branch of Fig. 3(a) shows that $W>W_h$ can occur either by setting-up the system to $T_o>T_h$ (human error) or may be due to the possibility that the value of the power corresponding to T_h in the look-up-table may be in error and actually corresponds to a higher temperature than T_h. The above second reason is also the primary event of the right branch.

Since the top event of Fig. 3(b) can occur during a cooling process, $W(T_o)$ is always smaller than or equal to W and the term $c(T_o-T)$ is negative or zero at equalibrium. Therefore $W(T_o)$ is solely responsible for the occurence of the top-event. $W(T_o)<W_l$ will occur if either $T_o<T_l$ (humen error) or if there is an error in the look up table, similarly as was described previously for Fig. 3(a).

The above discussion presents two kinds of faults, which may cause the system to exceed the temperature range.

1) A human operator error which causes the system to stabilize out of temperature range.
2) A "table error" which causes the power corresponding to T_l or T_h be lower or higher, respectively, than needed.

Both faults are caused by wrong input data to the control unit either by the look up table or by operator and have to be overcome. In order to do so the following steps have to be taken:

Step 1- The look up table must be carefully built with regard to the values of W_h and W_l.

Step 2- Hardware or software facilities should be used to limit the output power between W_l and W_h.

It should be noted that the second step intends to overcome the "human error". Generally such problems can be solved by using technical procedures. Hawever this is not recommended because human-beings are error prone.

DRASTIC CHANGE IN TEMPERATURE.

Figures 4(a) and 4(b) are a continuation of Fig. 2(b). It will be assumed that the faults described in the last paragraph were overcome. Hence the "in temperature range" situation will be discussed in this paragraph.

The top event in Fig. 4(a) is caused when the sum of $c(T_o-T)$ and $W(T_o)$ is greater than $W_f(T)$. this is true if either value is greater than $W_f(T)$ (left branch) or if both values are smaller than $W_f(T)$ but their sum is greater (right branch). these events are consequences of two primary events:

1) When the system is at temperature T it is set-up by an operator to T_o such that T_o-T is big enough ro cause an "over-power" supply. This is due either to the term $c(T_o-T)$ or to the difference between $W(T)$ and $W(T_o)$ or to both.
2) The difference between two consequent values of $W(T_o)$ in the look up table is great enpugh tp cause an "over-power" supply.

The top event of Fig. 4(b) may occur under the condition of a

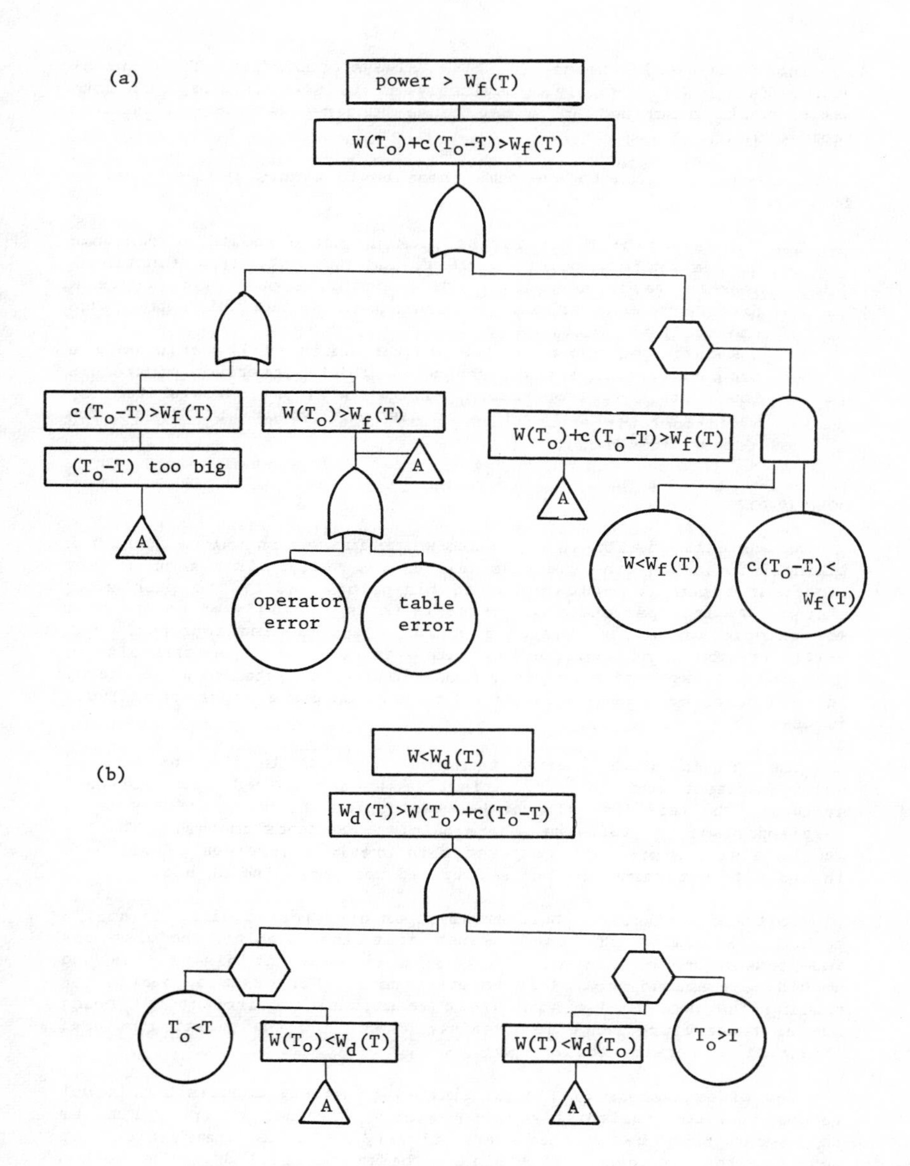

Fig. 4: (a) Drastic heating
(b) Drastic cooling

cooling process only. Hence $c(T_0-T)<0$ "always occurs" and the value of W depends on $W(T_0)$ solely. Obviously as is seen in Fig 4(b) both basic events described above may cause "under-power" supply causing rapid cooling.

A simple solution to overcome these basic evants is described as follows:

- The values of $W(T_0)$ in the look up table should be designed from a "safety point of view" and not only from functional aspects (e.g. accuracy). This design should not permit a difference between any two consequent values which may cause fault (2) to occur.
- Based on the above, a sub-program which will enable only a progressive temperature change should be inserted. Therefore the rate of temperature change will be decelerated by partically stabilizing on consequent values of T_0 in the look up table.

CONCLUSIONS

As was claimed, FTA is a process which focuses on unique scenarios taking into account the event as well as the state. This fact is very important since it enables one to pin-point exactly the problemtic states. An improper state of system is not necessarily an unsafe one. For example, an aircraft engine could be functioning improperly but still it would be considered "safe" as long as the aircraft is grounded. Risky states or risky conditions in a system might be hard, in some cases even impossible, to discover, unless a rigorous analysis is used.

The FTA is a well known technique for use in the analysis of safety systems and is very effective when combined with software systems. The fact that FTA can be used during the whole life-cycle is very important for software, where maintenance means changes. The FTA can be easily stored and compared with previous versions of software in order to determine whether safety had been affected or not.

Software systems and hardware systems are treated alike in an FTA method. By using FTA one assumes that the examined modules are independent of the others. This assumption is not always true and should be examined separately in every case. For example, heating or cooling the described systems is a fanctional requirement but being out of temperature range is a safety aspect and the relation between them must be considered.

The effectiveness of FTA is limited to safety aspects and is not recommended for reliability improvement. In order to apply FTA the top event must be defined very clearly. It is insufficient to describe the top event by saying, "The system failed". The method requires more accuracy! "The neutron flux in the nuclear reactor suddenly increased and control rods did not fall down". Other techniques are needed for systems which do not have the ability to determine catastrophic top-events, or which have a large number of critical safety failures.

ACKNOWLEDGMENTS

I am greatful to Dr. G. Leonard and Mr. E. Zell from the Israel Atomic Energy Commission for reading the first draft of this paper and for several helpful suggestions, and to Mrs. H. Crombine from the Soreq Nuclear Research Center for editing this paper.

REFERENCES

1. Rodgers, W. P., Introduction to System Safety Engineering. N.Y. Wiley, 1971.
2. Hammer, W., Handbook of Systems and Product Safety. Englwood Cliffs, N.J. Prentice-Hall, 1972.
3. Haasl, D. F., Roberts, N. N., Wesely, W. E. and Goldberg, F. R., Foult Tree Handbook. NUREG-0492, 1981.
4. Leveson, N. G. and Harvey, P. R., Analyzing Software Safety. IEEE Transactions on Software Engineering. 1983, 569.
5. Fryer, M. E., Risk Asssessment of Computer Controlled Systems. IEEE Transactions on Software Engineering. 1985, 125
6. Reifer, D. J., Software Failure Mode and Effect Analysis. IEEE Transactions on Reliability. 1979, 247
7. Deutsch, M. S., Software Veriffication and Validation. Prentice-Hall Inc. 1982.
8. Glass, R. L., Software Reliability Guidbook. Prentice-Hall Inc. 1979.

PROGRAMMABLE CONTROLLER FAULT TREE MODELS FOR USE IN NUCLEAR POWER PLANT RISK ASSESSMENTS

RAINA, V.M. and CASTALDO, P.V.

Design and Development Division - Generation
Ontario Hydro, Toronto, Canada

ABSTRACT

Programmable controllers are being used extensively in place of conventional relay logic in the design of Ontario Hydro's Darlington Nuclear Generating Station, a 4 x 850 MWe CANDU station due to achieve first criticality in November 1987. Among the systems controlled by programmable controllers are those that perform various safety functions. This paper describes the process followed to assess the impact of possible failure modes of the programmable controller on the safety of the plant. Some of the benefits realized from this process are described.

INTRODUCTION

Although digital computer control of major plant parameters such as reactivity, steam generator pressure and turbine power has been a standard feature of Ontario Hydro's nuclear generating units, the implementation of most other control logic has been relay-based. In the design of the Darlington Generating Station, however, such control logic is being realized by the use of programmable controllers (PC), instead of logic relays. Programmable logic controllers are being used not only in the control of normally operating equipment such as main boiler feedwater pumps, heavy water feed pumps and associated valves, but also to operate standby or emergency equipment, such as auxiliary boiler feedwater and emergency coolant injection pumps. The reliable performance of these devices is therefore of considerable importance.

Concurrent with the design activities for Darlington GS, Ontario Hydro has undertaken a probabilistic safety study in order to provide a quantitative estimate of the safety of the plant. This study involves the development of fault tree models for failures of major process and safety systems. Because of the extensive use of programmable controllers in these systems, it became necessary to develop fault tree models for various failures of the PCs.

In this paper, the development of programmable controller fault tree models is described. The paper first presents a brief description of the PC used, which is an Ontario Hydro-designed device called the OH-180, carrying out control and protective functions such as opening or closing pneumatic or motor-operated valves, starting or stopping pumps, or controlling circuit breakers in the power distribution system. Next, the development is presented of fault tree models that analyse in some detail the operation of each type of circuit board used in the PC, and the interactions between the circuit boards. Finally, the manner in which this information was used in determining how process system failure might occur due to failure of the PC is discussed.

OH-180 PROGRAMMABLE CONTROLLER

The OH-180 is a microprocessor-based programmable controller designed by Ontario Hydro specifically for the control of process systems in electrical power generating stations. Each PC comprises a CPU board, up to two power supply boards and input/output boards ranging in number from a minimum of one each to a maximum of eight, interconnected by means of a motherboard. The CPU board contains a CPU chip (Intel 8085A), 4x1K RAM (Intel 2114A), the electronic devices for communication between boards and the front panels, an RS232C interface, and a plug-in connector for a small board housing 6x4K EPROM (Intel 2732A) chips. The EPROM contains the instructions that enable the PC to carry out its control functions. External power supply to the PCs is obtained from the station 48 V dc power supply, which is battery-backed for increased reliability. The power supply board located in the PC provides a 5 V logic supply voltage and 48 V dc for contact sensing and driving output relays.

Each input board provides 24 two-terminal contact sense inputs. Two adjacent inputs can be configured, if desired by the user, as a Form C input to enable the PC to check for and alarm input contact discrepancy should one exist.

The output board has a total of 16 Form C contacts. Eight of these can be used to drive external loads using the PC's internal 48 V dc as the power source, and are therefore called driven outputs. The remaining eight are called dry outputs for use by external sensing circuits. The output board also contains circuitry to permit the state of the outputs to be sensed by the CPU so that a check may be made between the expected and the actual state of each output.

In addition to input and output discrepancy checks, the PC also locates ground faults occurring anywhere in the PC system or its associated field wiring, and carries out circuit continuity checks in each of the eight driven output circuits on an output board. The controller is also equipped with self-checking capabilities to monitor the health of major components such as the CPU, RAMs and EPROMs. In the event a serious fault is suspected, the PC causes the output relays to de-energize, thereby ensuring that the outputs attain a known failed state. Catastrophic failure of the CPU chip is detected by a watchdog circuit which de-energizes all outputs unless reset by a signal from the CPU.

The configuration of the PC and its application in a control system is as shown in Figure 1. The PC receives two-state information from devices located in the field, such as pressure switches and limit switches, in the main control room, such as handswitches, and in the various control equipment rooms, such as alarm units. It reads the state of the inputs every 10 ms, executes the program stored in the EPROM, and manipulates the output contacts as directed by the program. Typical end elements controlled by the outputs are motor starters, solenoid valves, circuit breakers, and indicating lamps and annunciators.

Figure 1: OH 180 Configuration

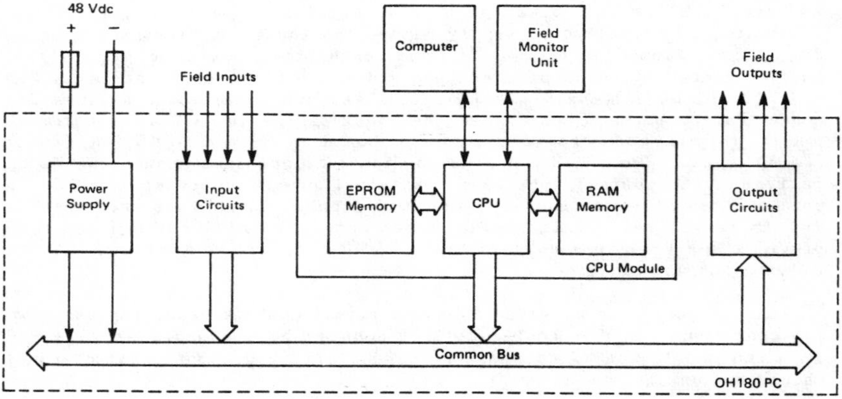

FAULT TREE DEVELOPMENT

The programmable controller carries out its functions in one or more of the following ways:

1. By opening or closing a normally-open or normally-closed, dry or driven, output contact on demand; or

2. By maintaining open or closed a normally-open or normally-closed dry or driven output contact as long as desired.

Failure analysis of the programmable controller, therefore, consists of examining ways and means by which it can fail to carry out the above functions. The PC failure modes of interest, thus, are as follows:

1. A normally-open or normally-closed, dry or driven output contact fails to open or close on demand;

2. A normally-open or normally-closed, dry or driven output contact spuriously opens (or fails to remain closed) or closes (or fails to remain open) during its mission.

A logical means of analyzing such failures is the fault tree technique. Accordingly, fault trees were developed with the above failure modes as top events.

The fault tree analysis was undertaken with the following objectives in mind:

1. Analysis of the programmable controller such that the reliability of the device itself could be estimated and a review carried out of its design; and

2. Development of a means to assess the impact of PC failures on the operation of the process system being controlled.

The former objective was met by developing and solving detailed fault trees for the top events of interest. The results of this activity were utilized in realizing the second objective by developing simplified models that could be conveniently incorporated into process system fault trees.

DETAILED FAULT TREE MODELS

A typical PC configuration comprising the CPU card and one each of the power supply, input and output cards was chosen for detailed analysis. The fault tree model development for a top event was carried out by starting from the output contact of interest and noting down the immediate causes of its failure. For example, the fault tree for the event 'Dry output contact fails to close on demand', shown in Figure 3, has as its immediate causes, with reference to Figure 2, the failing open of Resistor R93, failing open of connector A31, or the relay contact failing to close. The former two events relate to component failure and do not need further development since failure rate data are available at the resistor or connector level. Next, the ways in which the relay contact can fail to close were developed. The immediate causes are stuck open failure of the contact itself or failure to energize the relay. This is shown at the second level in the fault tree. Failure to energize the relay is, in turn, due to either failure of the relay itself or failure to provide a 48 V dc supply to the relay. The latter failure occurs if undetected loss of 48 V dc power to the output board takes place, or output driver U10 fails open undetected, or there is no signal to the driver. Failure of integrated circuit U10 is not developed further since failure rate data can be applied directly to this event. Among the causes of failing to provide a signal to the driver are failures of data and clocking signals to the latch U6.

The fault tree development proceeds in this manner until all output board failures that can affect the output contact of interest are analyzed, and one of the intermediate events is "Signal from the data bus on the motherboard incorrect". A manner in which this can occur is the existence of faults in those components on the CPU and other I/O circuit boards that are directly connected to the motherboard. The other mechanism is that the data signal to the motherboard from the CPU board is incorrect, which then entails an analysis of the CPU board.

Failure analysis of the CPU board was carried out in a similar manner by deductively identifying component failures that can affect the transmission of data, address and control signals to the output board, until failure of the microprocessor chip itself is postulated. Because most failures of the CPU chip, as well as the RAMs and EPROMs are detectable, the former by means of the watchdog circuit and the latter by the CPU, failure to detect and annunciate these faults was also analyzed. This process enabled identification of some failure modes which were not detectable in the original design.

Figure 2 Typical Output Circuit

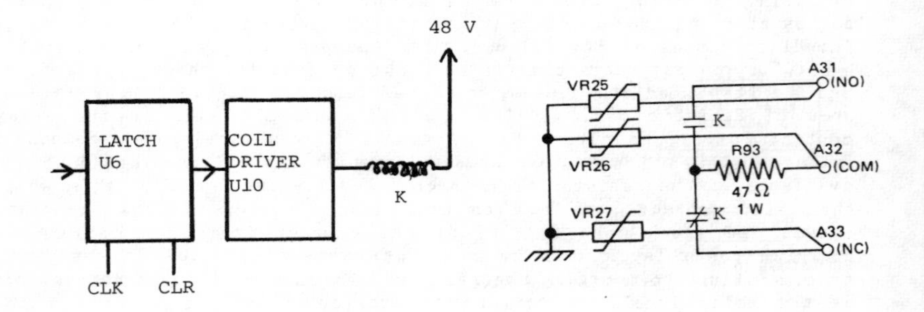

Figure 3 Fault Tree for Typical Output Circuit

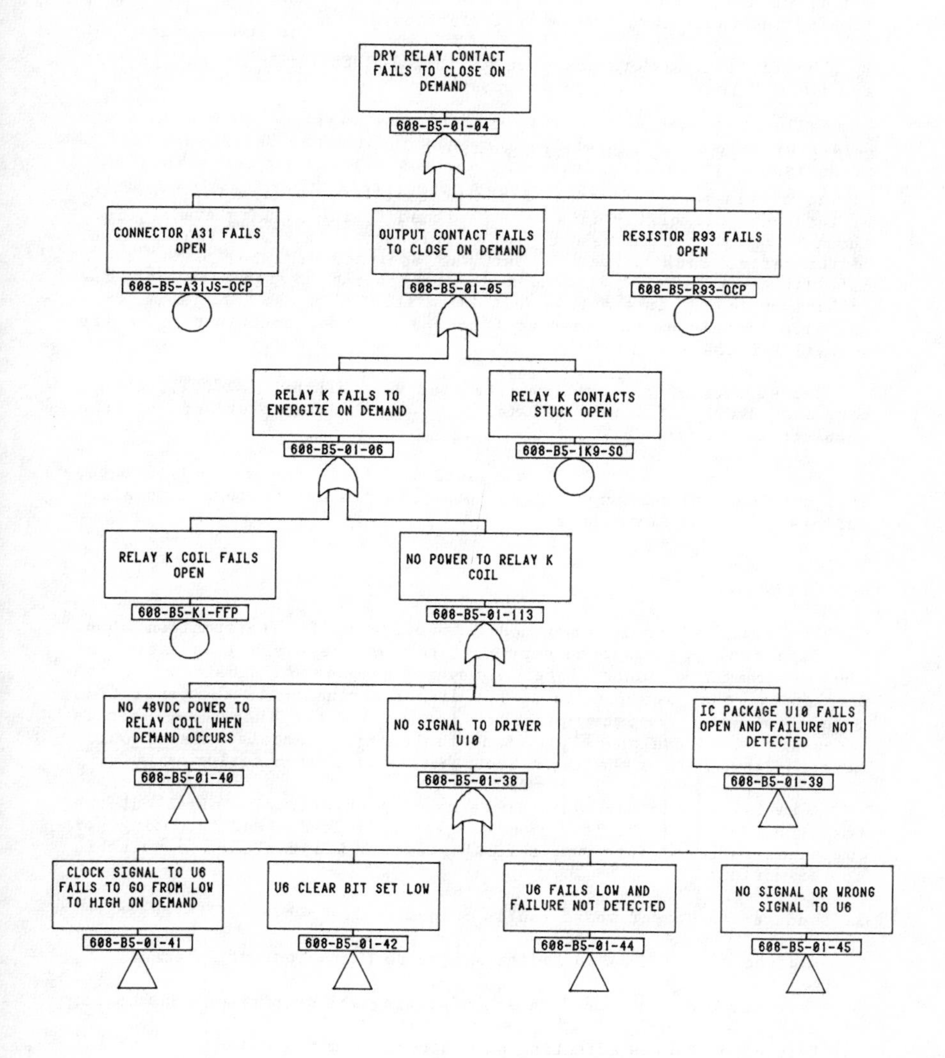

Finally, failure of input signals to the motherboard and hence the CPU, identified in the fault tree analysis of the CPU board, was analyzed. This led to the development of a fault tree for the input board, with the bottom events in the tree being the edge connectors linking the input card to the field devices.

The fault tree development also included failures of the 48 V dc and the 5 V logic supplies as appropriate.

RELIABILITY DATA AND FAULT TREE SOLUTION

To obtain some estimates of the probabilities of the failure modes of interest reliability data were assigned to the primary events, ie., those events not developed any further, in the fault tree. Additionally, each primary event was assigned a label based on a labelling scheme that included the component designation and the failure mode of interest. This permitted the use of fault tree solution techniques in order to find the dominant contributors to the overall failure probability.

The sources of reliability data used were MIL-HDBK 217D(1), except for the CPU, RAMs and EPROMs for which data provided by the manufacturer was used.

The computer code SETS(2) was used to obtain the minimal cutsets, ie., combinations of component failures containing the fewest possible failures, for the top events.

REDUCED MODELS

The detailed fault tree models provide useful information about the impact of various component failures on the overall operation of the programmable controller. However, because of their size and complexity, they are unsuitable for direct inclusion in the fault tree models of the various process systems. It, therefore, became necessary to develop simplified fault tree models that would, nevertheless, retain the major contributors to system failure.

A review of the dominant (in terms of probability) minimal cutsets for the detailed fault tree models indicated that grouping of component failures into the following categories would be a suitable approximation:

1. Undetected output board faults unique to the output of interest;

2. Undetected input board faults unique to the output of interest;

3. Undetected output board faults affecting all outputs on that board;

4. Undetected faults affecting all outputs from the PC;

5. Detected faults of types 1 to 4.

Accordingly, for each top event reduced fault tree models were developed whose primary events were of the types described in items 1 to 5 above. Failure probabilities for these were obtained from the minimal cutset lists for the detailed fault trees. In most cases, only some of the above categories of faults appear.

APPLICATION OF REDUCED MODELS

In analyzing process systems by means of fault trees, a point is reached at which erroneous operation of a field device due to faulty control signal is postulated. For example, a pump may fail to continue running due to its circuit breaker being demanded open erroneously by the control system. The topmost event in the fault tree of Figure 4, taken from a process system fault tree, shows such a failure mode. In this example, such an event can occur if two outputs, a driven one and a dry one, from a programmable controller close spuriously. Figure 4 also shows the development of one of these events by making use of one of the reduced models. All control system failure events in a process system fault tree are similarly developed, following which minimal cutsets for the process system fault tree are obtained. This permits assessment of the common mode failure potential of the programmable controller should there be any.

Figure 4 Example of A Reduced PC Model

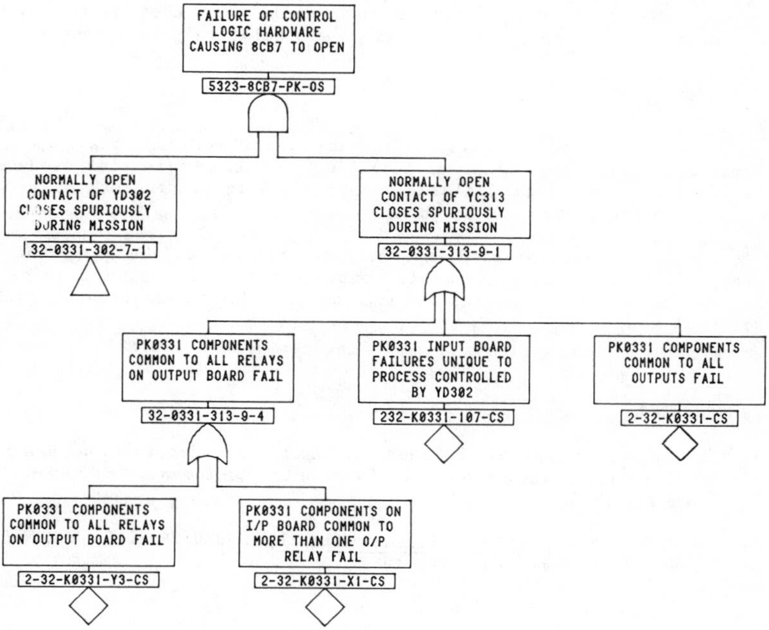

At the present time, programmable controller models have been developed in the study for control faults identified in most process system fault trees, thereby enabling quantification of the impact of PC failure modes on the reliable operation of the system. When all such models are incorporated into their respective system fault trees, probabilities of occurrence will be calculated of those combinations of system failures that affect the public health risk from the station.

BENEFITS

The detailed fault tree model development identified a number of areas where reliability improvements were possible. These were mainly in the area of providing better fault detection capability and taking appropriate action such that the outputs were left in the de-energized, or 'safe', state following the occurrence of the fault. It was found possible to implement these improvements mostly by means of software changes to the executive program.

In addition, the review of programmable controller implementation of process control logic by means of fault tree models revealed situations wherein the assignment of a number of control functions to the same programmable controller led to some loss of system redundancy. These were corrected by re-allocation of the control functions to different PCs.

Because the above-noted insights into system design were available while the generating station was still in the design phase, incorporation of design modifications was relatively inexpensive.

CONCLUSIONS

The fault tree modelling process described in this paper has provided a practical means of including in large-scale risk assessment studies the effect of failures of programmable controllers on process systems being controlled by them.

It has also enabled a thorough review to be made of the reliability of the programmable controller itself, as well as of the manner in which the device is utilized in the realization of process control functions.

REFERENCES

1. U.S. Department of Defense, Military Standardization Handbook: Reliability Prediction of Electronic Equipment, MIL-HDBK-217D, January 1982.

2. Worrell, R.B., SETS Reference Manual, NUREG/CR-4213, US Nuclear Regulatory Commission, May 1985.

THE INTEGRATED PROTECTION SYSTEM: HIGH INTEGRITY DESIGN AS A RESPONSE TO SAFETY ISSUES

R. CIPRIANI, F. MANZO, F. PIAZZA

ANSALDO S.p.A./Div. NIRA
Via dei Pescatori, 35
GENOA, Italy

ABSTRACT

The major functions and architecture of a Nuclear Power Plant Protection System are designed to meet the safety issues. Both the analogic and the digital designs are able to satisfy the requirements.
A computer based architecture presents a higher integration capability and is characterized by distributability, modularity and flexibility features. To face faults, errors and other anomalous or unexpected behaviours, fault-tolerance and self diagnosis can be implemented. WESTINGHOUSE, ANSALDO S.p.A./Div. NIRA and various Italian organizations are developing, under WESTINGHOUSE responsibility, a microprocessor based Integrated Protection System (IPS), whose high integrity design is presented in the paper.

INTRODUCTION

The safety of a Nuclear Power Plant relies on a set of automatic actions that on most of the plants are implemented within analog process protection equipment. This equipment, designed in the 60s and early 70s, receives inputs from sensors, provides information to the operator, performs calculations on these values and compares the results to allowable limits. If the limits are exceeded a partial reactor trip is generated. External logic performs a voting algorithm on the partial trips from the redundant protection sets and conditionally generates a reactor trip. A similar path exists for the generation of Engineered Safeguard Systems actuation. These actuations mitigate the effects of an undesired event.
As shown in fig. 1 these protection channels need to interface different subsystems, based on different technologies (solid state, relays, dynamic logic).

The limited processing capabilities of these technologies allow to reach an adequate reliability only limiting the complexity of the functions. For that reason only the direct safety functions could be properly addressed by the design, while several aspects, that could indirectly impact on safety, remained as potential areas of concern.

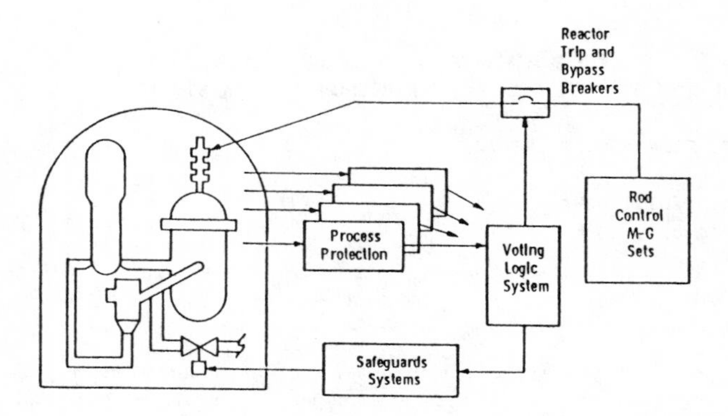

Figure 1: Protection System

Nowaday the maturity of the microprocessor technology allows to implement much more complex functions within system configurations relatively simple. This means that the design can be developed at a very high level of integrity addressing, since the early phase, each of the safety issues in the most proper and reliable way.
The Integrated Protection System (IPS) is a project that ANSALDO div. NIRA is developing since Jan. 84 jointly with Westinghouse El. Co. to be used as the main protection system for the PWR's planned by the Italian Energy Plan, the first two units of which were ordered to ANSALDO in Dec. 85.
This presentation wants to be an overview of the main design goals identified as answers to the main safety issues and of the design choices adopted to meet those goals.

SYSTEM RELIABILITY

The highest system reliability as possible, intended as the lowest probability to fail on demand, is of course the main design goal from a safety point of view. In conventional systems the reliability issue was usually addressed making the system tolerant to random failures through redundancy.A much more comprehensive approach was assumed for the IPS design in consideration that was not possible to identify a unique answer to the reliability issue. The following design criteria were established for the development:

- Redundancy: in addition to the typical four redundant channel sets configuration and 2/4 voting logic the IPS provides extensive redundancy within each channel set that allows to maintain complete operability and integrity of the system even in presence of maintenance or failures.
- Functional diversity: alternative and independent parameters are monitored by the IPS to protect the plants against the Design Basis Events.
- Defense-in-depth: the system functions are rigorously organized into different echelons of defense of the plant.
- Independence: in addition to an improvement of the separation between redundant channel sets through an extensive use of fiber optic communications, complete functional and electrical independence is provided within each channel set between sub-systems implementing functionally diverse actions or functions at different echelons of defense. The integration of functional diversity, defense-in-depth and independence makes the system extremely tolerant even to common mode failures.
- Fail sale design: the typical dynamic operation of the microprocessor technology allow to detect most of the malfunctions through internal self-diagnostic and to take fail safe actions in case of failure.

  In critical areas, such as the reactor trip bus, extremely reliable and intrinsecally fail safe technology are used (dynamic magnetic logics).
- Testability: built-in automatic test sub-systems are provided within each channel set to allow the complete test of a channel set in approximately one hour. This dramatic reduction of the testing time allow to reduce the test interval with a direct improvement of the reliability figures.

IMPROVED AVAILABILITY

In conventional systems the design criteria, stated to improve the reliability, usually penalized the system availability. In addition to the obvious economical effect due to undesired plant outages, the high probability to have spurious actuation had a negative impact on safety as an intervention of the safety systems must always be considered as an undesired challange to the plant.
The IPS design optimizes the reliability and availability targets through the following design criteria:

- By-passes: a sophisticated logic is provided that allow to by-pass each channel or each channel set in case of test or maintenance, and to convert the voting logic in the remaining operating channels from 2/4 to 2/3;
- Improved safety limits: the microprocessor processing capabilities

allow to monitor, on the basis of complex algorithms, directly the core limits (DNBR and KW/ft) instead of the old design that monitored indirect parameters (overpower and overtemperature).

- Redundancy: the redundancy provided within each channel set allows complete on-line testing with complete availability of the system and of the plant.
- Automatic tests: the reduced time required for testing minimizes the time of partial unavailability of part of the system.
- System monitoring: the possibility to easily communicate via optical data-link the entire system status allows the operator to have a complete monitoring of the system performance.

ADVANCED TECHNOLOGY

In consideration of the construction time of a Nuclear Plant (about 10 years) and of the rapid development of the electronic technology, the use of the most advanced mature technology is essential to guarantee to maintain the system throughout its entire life cycle. In addition some technological choices are a direct answer to some safety issues:

- System integration: the microprocessor capabilities allow the integration within the same system and the same technology of the entire safety functions, from the sensors to the actuators. This drastic semplification of the interfaces minimizes the probability of a lot of very common errors;
- Fiber optics: the extensive use of fiber optics for communication within the system, in addition to improve the separation between redundant parts, is a fundamental choice to make the system immune to the effects of plant induced electrical and/or electromechanical interferences.
- Flexibility for modification: one of the typical unresolved safety issues for the conventional systems is the complexity to incorporate changes during the system life cycle.
  The distributed architecture of the IPS is such that most of the predictable changes can be incorporated only through very easy software modifications.
- Use of high level programming languages: the availability of very reliable high level structured languages (PL/M-86) in conjunction with the use of commercial available hardware (INTEL SBC Multibus compatible hardware) allows to organize the project development with the structure that minimize the traditional interfaces between the plant engineer, the system designer and the manufacturer that are usually one of the main sources of errors.

EASIER MAINTENANCE

The test and maintenance of the conventional protection systems are not only a problem from a point of view of the plant operation, but are also a typical safety issue for the relevant probability of errors due to complex and/or improper procedures, long test and maintenance intervals, complex personnel skillness both on plant operation and electronics. The IPS design address all those issues trying to make the system test and maintenance as easy as possible through the following characteristics:

- Automatic test: the replacement of procedurized tests with automatic periodic tests, in addition to reduce the effort requested and semplify the procedures, is an improvement of the completeness and effectiveness of the tests that are not affected by personnel errors.
- Self diagnostic: extensive self diagnostic permits to locate faults quickly minimizing the possibility of errors and the diagnosis effort. Typical errors that are automatically and continuously checked are:
  RAM Check, PROM Checksum, APU Checks, NURAM Checks, Deadman Timer, Sensor Checking, Output actuator circuitry integrity.
- Modular replacement: the design is such that the faults repair is carried out on-line by replacement of modular devices with no degradation of the system functions other than those associated to the faulted equipment. This reduces the possibility of errors and the MTTR with a direct improvement of the system reliability.
- System calibration: the system calibration is in "natural" engineering units minimizing the usual source of errors due to the conversion from engineering units to electrical units and to pot set.
- On-line test and maintenance: all the test and maintenance procedures can be carried out on line with no need to modify the system configuration.

It must be noted that the system fault tolerant design reduces sensitivity to the maintenance errors.

EASIER INSTALLATION

Electrical installation is usually a major source of cost and more of risk of delays in plant construction and commissioning. In addition the installation remains a safety issues for the difficulty to find always satisfactory solutions for separation and protection.

The IPS design, allowing a dramatic reduction of the interconnection, can positively solve the installation problems together with a very significant cost savings.

- Distributed architecture: the IPS architecture is designed to allow a distributed layout for those cabinets, such as the component interposing logic, that have a heavy interconnection to the field. This distributed layout reduces connection length, reducing directly one of the major sources of induced noise and interferences.

- Multiplexing: extensive use of signal multiplexing, in conjunction with the distributed architecture, reduces conductor count in congested areas (for example: Control Board, Spreading Areas). This reduction simplifies the problem to get good separation between different classes of cables and between redundant parts.
- Integration and uniformity: the total system design integration and the use of an uniform technology and cabinetry simplify the installation and yield power interfaces problems.

HIGH QUALITY OF DESIGN

As discussed above the IPS, through its high integrity design approach, addresses most of the typical safety issues. Of course the consequence is that the system design is characterized by a significant functional complexity and by a very large content of software. This means that new and specific issues must be addressed mainly related to the control of a complex design process and to the software generation.

A design developed at the highest standard of quality is the only answer to those issues.

- Software generation: all the software is produced according to rigorous standard and procedures. Top-down structured architecture, rigid programming rules, guidelines to discourage tricky programming solutions are stated in order to avoid unpredictable operation of the software.
- Verification and Validation: independent verification at system, hardware and software level is carried out through the entire design process at each development phase to guarantee the consistency with the requirements stated at the previous phase.
  A complete integrated verification environment was established including tools for automatic test cases generation and test reports production.
  A final system validation is planned to be carried out connecting the system to a plant simulator and verifying that the system performance against the Design Basis transients is consistent with a correct implementation of the functional requirements.
- Configuration Management: the entire software life cycle is kept under a rigid Configuration Management control to guarantee the consistency of the software with the system design.
- Documentation structure: since in the early stage of the project it was recognized that in order to reach a real effectiveness of all the above methodologies and guidelines, a rational organization of the design activities and a well defined documentation structure and standardization was needed. The System Development and Implementation Process (SYSDIP) is the program that was established as the answer to that issue. In fig. 2 the SYSDIP Logic is outlined. As

show in the diagram the entire development process is structured into the following phases, each one complete and consistent at its own level of detail:

- System requirements
- System design
- Hardware and Software sub-systems and module design
- Hardware and Software sub-systems and modules implementation
- System integration.

The independent verification loops are identified for each design phase.

The documentation to be produced at each phase is also identified and stadardized defining for each type of document the standard index and the typical content of each paragraph.

SYSTEM DESCRIPTION

Fig. 3 shows a general block diagram of the IPS architecture. The system consists of three major sub-systems:

- Integrated Protection Cabinets (IPC's)
- Engineered Safety Features Actuation Cabinets (ESFAC's)
- Integrated Logic Cabinets (ILC's).

The IPC's consist of four redundant channel sets each one implementing the following major functions:

- to interface the process sensors through signal conditioning boards;
- to provide calculations to generate partial trip signals both for reactor trip and ESF;
- to communicate reactor partial trip signals with the other channel sets;
- to vote the partial trip signals and to generate reactor trip signal for the associated set of breakers;
- to communicate ESF partial trip signals to the ESFAC;
- to automatically test the entire channel set;
- to communicate information to external system as the Plant Computer.

The ESFAC's are sub-systems whose level of redundancy is plant dependent as each is associated to a fluid system train (four train are provided for the Italian Plant). Each ESFAC implements the following major functions:

- to receive ESF partial trip signals from the four IPC's;
- to vote the partial trips and to generate system level actuation signals for the ESF;
- to communicate the system actuation signals over a redundant optical data highway;
- to automatically test the entire sub-system;
- to communicate information to external systems.

The ILC's are modular cabinets dedicated to implement component level logic for process actuators as pumps, valves, fans, etc.. Each ILC interfaces over the redundant data highway through which can receive

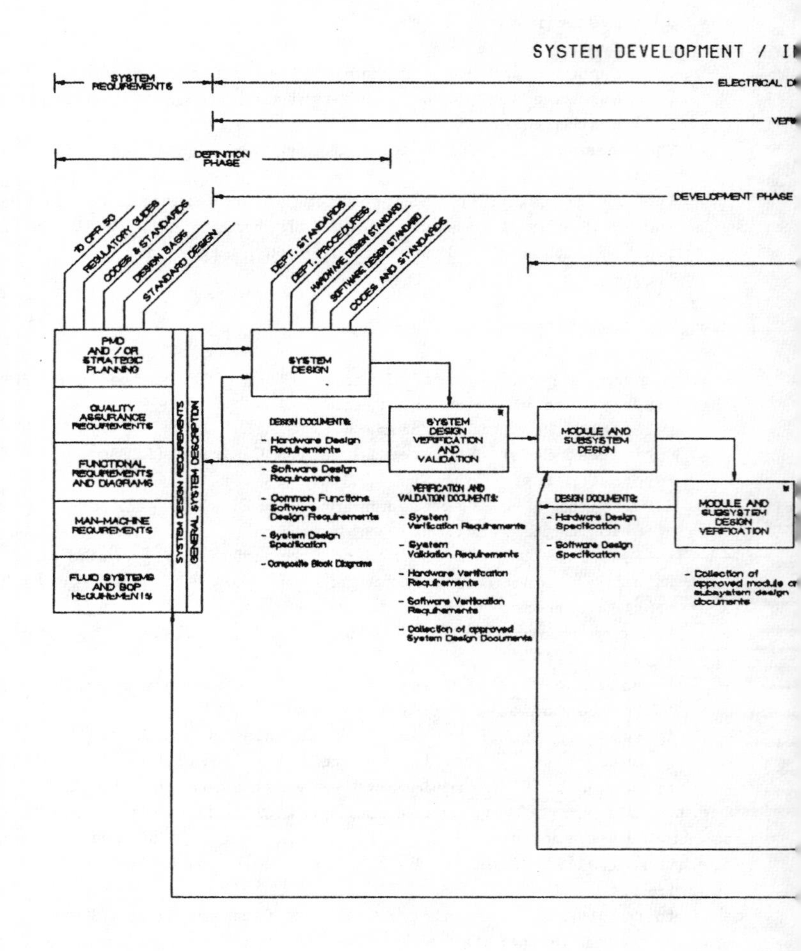

SYSTEM DEVELOPMENT / I
SYSTEM REQUIREMENTS
ELECTRICAL D
DEFINITION PHASE
DEVELOPMENT PHASE
10 CFR 50
REGULATORY GUIDES
CODES & STANDARDS
DESIGN BASIS
STANDARD DESIGN
DEPT. STANDARDS
DEPT. PROCEDURES
HARDWARE DESIGN STANDARD
SOFTWARE DESIGN STANDARD
CODES AND STANDARDS
PMD AND / OR STRATEGIC PLANNING
QUALITY ASSURANCE REQUIREMENTS
FUNCTIONAL REQUIREMENTS AND DIAGRAMS
MAN-MACHINE REQUIREMENTS
FLUID SYSTEMS AND BOP REQUIREMENTS
SYSTEM DESIGN REQUIREMENTS
GENERAL SYSTEM DESCRIPTION
SYSTEM DESIGN
DESIGN DOCUMENTS:
- Hardware Design Requirements
- Software Design Requirements
- Common Functions Software Design Requirements
- System Design Specification
- Composite Block Diagrams
SYSTEM DESIGN VERIFICATION AND VALIDATION
VERIFICATION AND VALIDATION DOCUMENTS:
- System Verification Requirements
- System Validation Requirements
- Hardware Verification Requirements
- Software Verification Requirements
- Collection of approved System Design Documents
MODULE AND SUBSYSTEM DESIGN
DESIGN DOCUMENTS:
- Hardware Design Specification
- Software Design Specification
MODULE AND SUBSYSTEM DESIGN VERIFICATION
- Collection of approved module or subsystem design documents
* VERIFICATION REQUIRED FOR SAFETY SYSTEMS ONLY

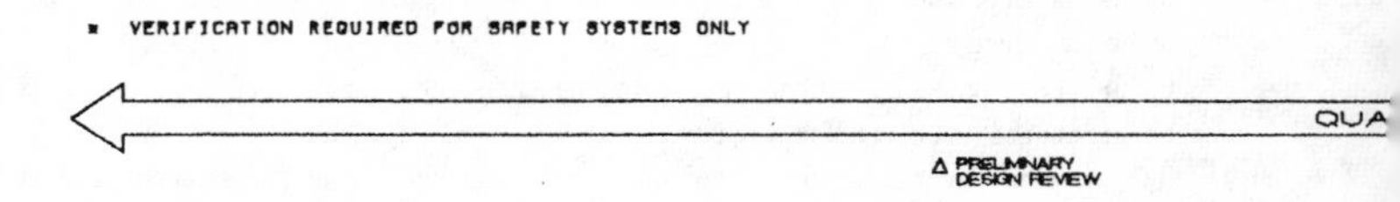

QUA
Δ PRELIMINARY DESIGN REVIEW

FIGURE 2

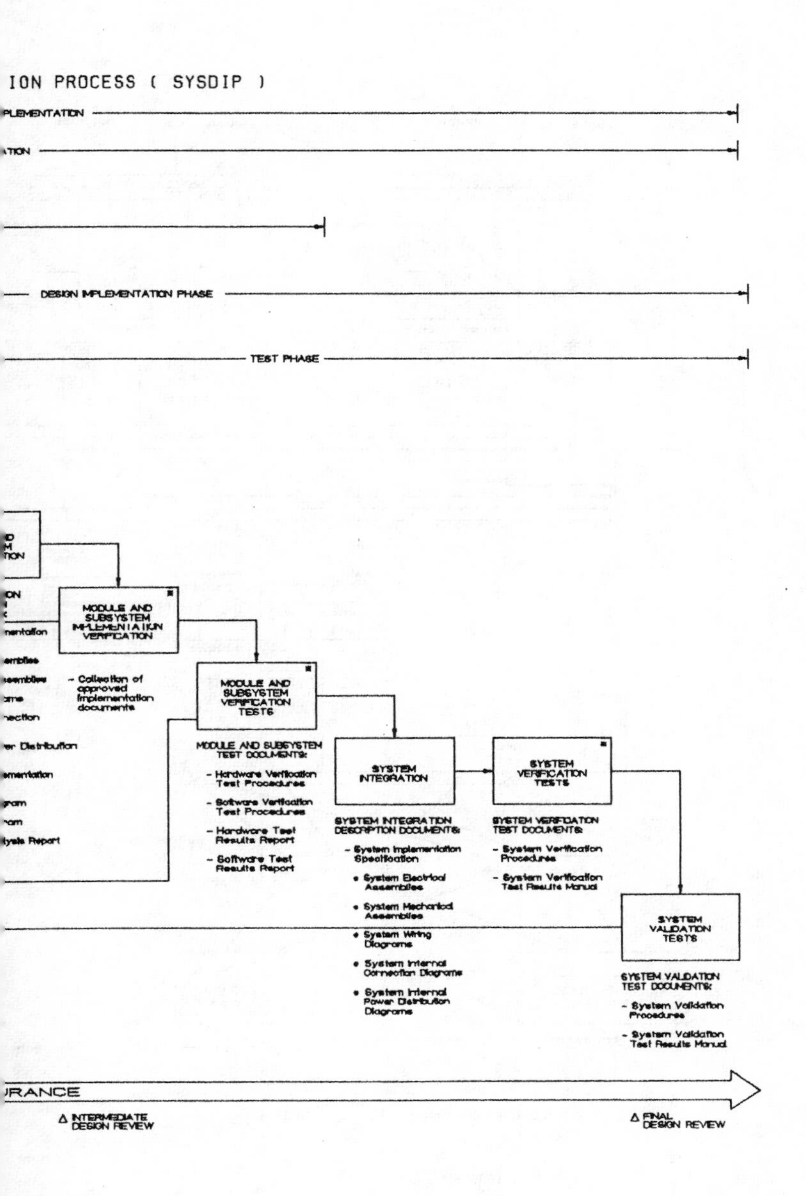
ION PROCESS (SYSDIP)
DESIGN IMPLEMENTATION PHASE
TEST PHASE
MODULE AND SUBSYSTEM IMPLEMENTATION VERIFICATION
- Collection of approved implementation documents
MODULE AND SUBSYSTEM VERIFICATION TESTS
MODULE AND SUBSYSTEM TEST DOCUMENTS:
- Hardware Verification Test Procedures
- Software Verification Test Procedures
- Hardware Test Results Report
- Software Test Results Report
SYSTEM INTEGRATION
SYSTEM INTEGRATION DESCRIPTION DOCUMENTS:
- System Implementation Specification
• System Electrical Assemblies
• System Mechanical Assemblies
• System Wiring Diagrams
• System Internal Connection Diagrams
• System Internal Power Distribution Diagrams
SYSTEM VERIFICATION TESTS
SYSTEM VERIFICATION TEST DOCUMENTS:
- System Verification Procedures
- System Verification Test Results Manual
SYSTEM VALIDATION TESTS
SYSTEM VALIDATION TEST DOCUMENTS:
- System Validation Procedures
- System Validation Test Results Manual
URANCE
Δ INTERMEDIATE DESIGN REVIEW
Δ FINAL DESIGN REVIEW

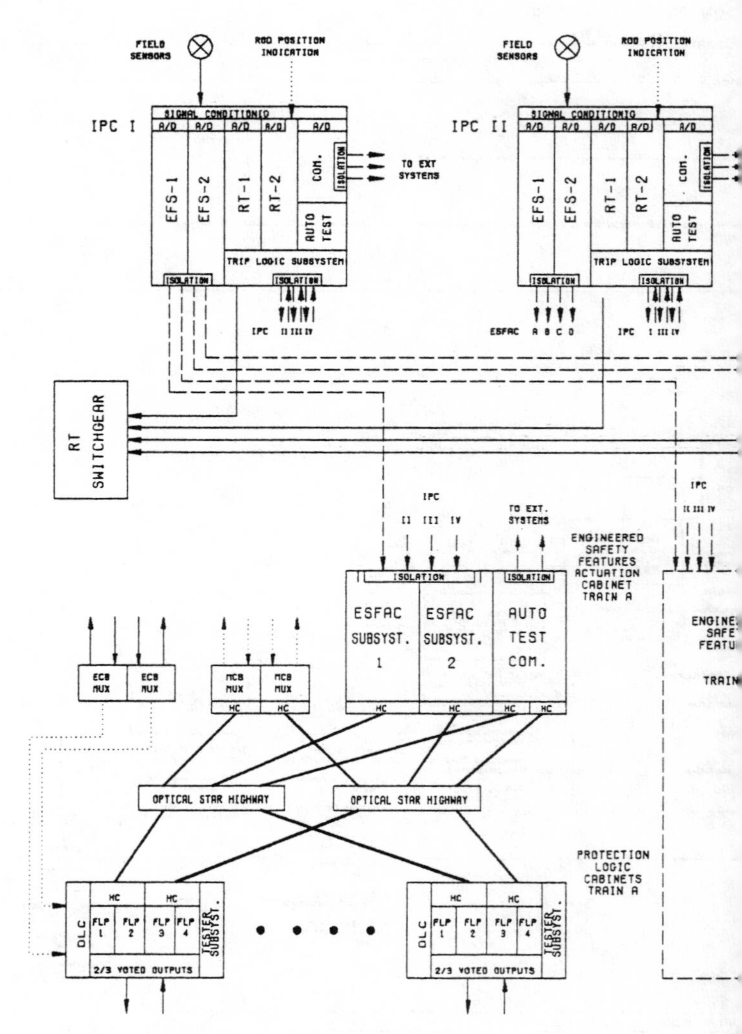
FIELD SENSORS
ROD POSITION INDICATION
IPC I
SIGNAL CONDITIONING
A/D
EFS-1
EFS-2
RT-1
RT-2
COM.
ISOLATION
TO EXT SYSTEMS
AUTO TEST
TRIP LOGIC SUBSYSTEM
IPC II III IV
IPC II
ESFAC A B C D
IPC I III IV
RT SWITCHGEAR
IPC
II III IV
TO EXT. SYSTEMS
ENGINEERED SAFETY FEATURES ACTUATION CABINET TRAIN A
ESFAC SUBSYST. 1
ESFAC SUBSYST. 2
AUTO TEST COM.
HC
ECB MUX
MCB MUX
OPTICAL STAR HIGHWAY
PROTECTION LOGIC CABINETS TRAIN A
DLC
FLP 1
FLP 2
FLP 3
FLP 4
TESTER SUBSYST.
2/3 VOTED OUTPUTS
MOTOR CONTROL CENTERS, VALVES, ...- TRAIN A

FIGURE 3

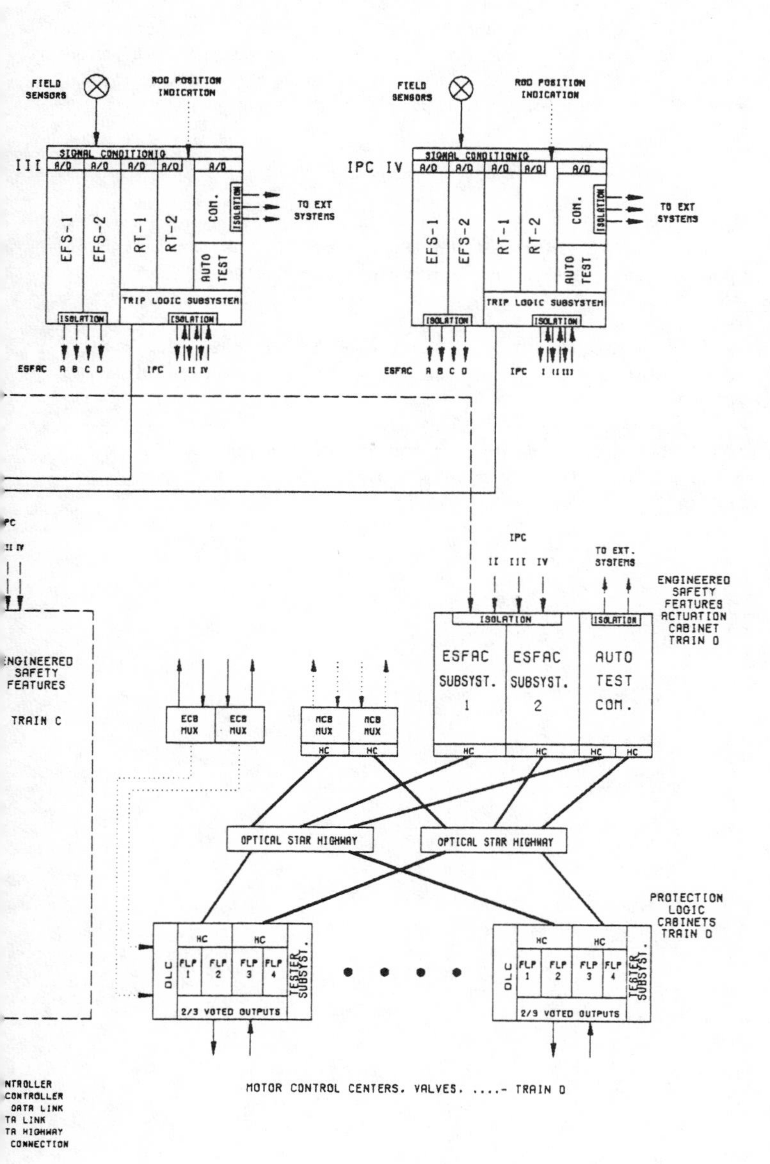
FIELD SENSORS
ROD POSITION INDICATION
SIGNAL CONDITIONING
III
A/D
EFS-1
EFS-2
RT-1
RT-2
COM.
ISOLATION
AUTO TEST
TO EXT SYSTEMS
TRIP LOGIC SUBSYSTEM
ISOLATION
ESFAC A B C D
IPC I II IV
IPC IV
IPC I II III
IPC
II III IV
TO EXT. SYSTEMS
ENGINEERED SAFETY FEATURES ACTUATION CABINET TRAIN D
ESFAC SUBSYST. 1
ESFAC SUBSYST. 2
AUTO TEST COM.
HC
ENGINEERED SAFETY FEATURES
TRAIN C
ECB MUX
MCB MUX
OPTICAL STAR HIGHWAY
PROTECTION LOGIC CABINETS TRAIN D
OLC
FLP 1
FLP 2
FLP 3
FLP 4
TESTER SUBSYST.
2/3 VOTED OUTPUTS
MOTOR CONTROL CENTERS. VALVES. ....- TRAIN D
NTROLLER
CONTROLLER
DATA LINK
TA LINK
TA HIGHWAY
CONNECTION

actuation signals either from the ESFAC (automatic actuation) or from a MUX interfacing the Main Control Board (manual actuation).
The status of the actuated equipment is transmitted by the ILC's over the data highway to be displayed on the Main Control Board. The actuation logic is implemented by three redundant microprocessors and the actuation signals are voted in a 2/3 Logic in the power interface boards. Built-in automatic tester is provided for each ILC.

REFERENCES

1. GALLANGER J.N., IAEA Guidelines and IEC recomendations for design of nuclear power plant control and instrumentation systems. IAWA-SM-265

2. S. BAGNASCO, F. MANZO, F. PIAZZA, Requirements and sesign for a distributed computerized system for safety and control applications. IFAC SAFECOMP 85

3. D. M. RAO ,S. BOLOGNA, Verification and Validation program for a distributed computer system for safety application. IFAC SAFECOMP 85

ENHANCING SYSTEM RELIABILITY BY IMPROVING COMPONENT RELIABILITY

KLEPPMANN, WILHELM G.

TÜV Stuttgart e.V.
IC-Test, Prüflabor für Elektronik
Gottlieb-Daimler-Str. 7, D - 7024 Filderstadt

ABSTRACT

The reliability of systems can be increased by using redundancy and/or self-testing systems. However, given a certain structure of the system, its reliability and availability can also be improved considerably by increasing the reliability of its individual components. The following measures lead to an increase of the reliability of electronic components:

- controlled environment
- derating and safety margins
- qualification of the components for the proposed environment
- detailled incoming inspection
- burn-in and other screening measures

The efficiency of these measures is discussed and practical examples are given.

INTRODUCTION

If programmable electronic systems are to be used in safety systems, a high reliability is required. The detailled reliability goals depend on the application in consideration.

Since in any system failures can occur, it is generally required that in safety systems single failures must not lead to critical conditions. Depending on the application it may also be required that certain combinations of failures do not lead to critical conditions and/or that failures are recognized immediately after their occurrence, e.g. by self-test, so that countermeasures can be taken.I.e. redundancy and self-test are required. The detailled requirements depend on the application in consideration and also vary from country to country. They are the subject of other papers at this Symposium and will not be dealt with here.

The purpose of this paper is to describe some methods to further improve the reliability of electronic systems, given that all the basic requirements, such as redundancy, self-test, correctness of design etc. are fulfilled. In the following several methods for enhancing the reliability of electronic systems by improving the reliability of the individual components are by presented and discussed.

Their purpose is not to replace basic requirements such as redundancy, but rather to supplement them.

CONTROLLED ENVIRONMENT

The reliability of electronic components depends an the severity of the environmental conditions to which they are subjected during use. Influences such as

- high temperatures
- high humidity
- corrosive gases
- water and salts
- temperature variations
- vibration and mechanical shock etc.

lead to an increase of the failure rate of the components and thereby to a decrease of the reliability of the electronic system. MIL-HDBK 217 (1) quantifies the influence of the enviroment on the failure rate by an enviromental factor with which the basic failure rate has to be multiplied. This factor varies from 0.38 for use in climatized rooms to 220 for rocket. launch, i.e. the failure rate varies by a factor of 540 from the most benign to the harshest.In addition to that many failures are accelerated by high temperatures. The failure rate depends on the junction temperature via the activation energy.The activation energy depends on the failure mechanism. For most mechanisms it lies in the range 0.3 eV to 1.0 eV (2). An increase of the junction temperature eg. from 50°C to 100°C leads to an increase of the failure rate

by a factor of 4 for an activation energy of 0.3 eV

by a factor of 120 for an activation energy of 1.0 eV.

Control of the environment will therefore lead to lower failure rates of electronic equipment and increase its reliablility. Unfortunately the enviromental conditions cannot always be controlled. However, frequently one can reduce the severity of environmental conditions by measures such as

- housing electronic equipment in special rooms or racks away from the process to be controlled
- forced convection or climatization.

DERATING AND SAFETY MARGINS

Derating is the reduction of electrical, thermal and mechanical stresses of a component. The aim of derating is the reduction of the speed of aging and wear-out processes. Generally aging processes slow down much more than linearly with the reduction of stress levels. Therefore derating can lead to a marked reduction of failure rates and an increase of useful lifetime.Examples for derating are the use of

- CMOS-ICs at less than the maximum rated supply voltage
- resistors at reduced power
- capacitors at reduced voltage.

By careful dimensioning and safety margins drift effects to be expected can be allowed for and failures of the system can thus be avoided. The drift may be caused by aging or by adverse environmental conditions such as too high temperature. Since large drifts are less likely or take longer than small drifts, conservative dimensioning can reduce the overall failure rate of the system and increase its useful lifetime.

Derating and safety margins are for example required by the European Space Agency (3).

QUALIFICATION OF THE COMPONENTS

All components of a system have to be suitable for the environmental and electrical conditions to be expected. The suitability of a component has two facets:

- Even under the most adverse imaginable conditions during the proposed application the function must still be guaranteed and all relevant parameters must be within specified limits.
- Aging phenomena under the expected conditions must be sufficiently slow.

In qualification tests both these aspects are examined. A sufficiently large sample of components is electrically characterized at different temperatures and subjected to aging stresses. The aim is to get information on the whole lot from which the sample was taken. This is particularly important when new types of component are to be used, since in MIL-HDBK 217 (1) and in Electronic Reliability Data (4) for new components a learning factor of 10 is given. That means that for newly developped components the average failure rate is by a factor of 10 larger than for equivalent components from mature production. This is the result of some new component types with design faults or initial manufacturing problems. The aim of qualification tests is also to discover these problematic component types.

In the electrical characterization the electrical parameters and the function of a component under extreme conditions are measured on a sufficient number of components. In this way the behaviour of the component under extreme conditions becomes known. Parameters for the proper dimensioning of systems can be obtained, if nonspecified parameters are important. Also it can be checked, whether all specified parameters remain within their limits.The behaviour of components and system under extreme conditions is particularly important for safety systems, since frequently these systems are only needed to control accidents i.e. just when the environmental conditions are extreme.

In the stress part of a qualification test the components are subjected to extreme environmental conditions over a longer period of time, to check whether these conditions lead to parameter drift, aging and failures of components with time. By the extreme environmental conditions aging phenomena are accelerated so that drifts and failures occur after a reasonably short time. However, it is important that no failure mechanisms are activated which would not occur during normal operating conditions. Qualification procedures are described in detail e.g. in IEC-68 (5), CECC (6), MIL STD 883 (7) and in the review article (8).

INCOMING INSPECTION

In incoming inspection the function, all specified parameters and, if necessary for the particular application, additional parameters are measured. If any parameters are outside the specified limits or the specific limits for the particular application the component is rejected.

By incoming inspection faulty components are recognized before they are soldered onto the board. This reduces the amount of fault-finding on board level and repair of boards. Since board-repair is expensive, incoming inspection may reduce the overall production cost.

However, in this context, reliability aspects are more important:

- Board-repair may damage the board, other components or solder points.
- A board may function, even if parameters of individual components are outside their specified limits, particularly if safety margins are used in design. But parameters outside the specification mean that there is less margin for drift left. Also they are frequently an indication of other problems which lead to early failures.

In other words, sorting out failed components and components with parameters outside the specification before board assembly leads to an increase of the reliability of the boards.

BURN-IN AND OTHER SCREENING MEASURES

Even amongst components which lie within the specifications there are usually some weak items. The weak components will generally fail early during field use, typically during the first months of operation and give rise to an increased failure rate during that period (early failures). These weak components cannot be recognised by an electrical test alone, since during the test they are operated only for a few seconds.

The aim of burn-in and other screening methods is to trigger failures of most of the weak components without impairing the useful life of the main population of good components. In that way the number of early failures is decreased noticeably.

Burn-in is the operation of the components at elevated temperatures (typically 125°C) for an extended time (typically 24-240 h).Burn-in is suitable for accelerating most chip-related failure mechanisms. The temperature has to be chosen so that it leads to a sufficient acceleration of the normal failure mechanisms without introducing new mechanisms.

An example for other screening methods is temperature cycling, which activates many package-related failure mechanisms. Details of these screening methods and other methods are given in the references (5-8).

Screening is always followed by an electrical test to find the failed components, i.e. incoming inspection is part of screening. A statistical evaluation of failure rates with and without screening shows that a lowering of the average failure rate by a factor of about 3 can be achieved by screening (1,4). This means that screening is an efficient way of increasing the reliability of electronic components.

CONCLUSIONS

Several methods have been described, by which the reliability of individual electronic components can be increased. With the reliability of the components, the reliability of the system as a whole increases. Risks related with the use of programmable electronic systems in safety applications can be reduced by these methods. Since reliability also increases customer satisfaction and reduces repair and warranty costs, these are suitable voluntary measures of the manufacturer to further increase the reliability of electronic systems. In some cases the application of these methods is also required by technical rules or it is a contractual requirement by the customer.

REFERENCES

1. Military Handbook 217 D: Reliability Prediction of Electronic Equipment, US Department of Defense, 1982

2. Peck, D.S., Trapp, O.D., Accelerated Testing Handbook, Technology Associates, 1981

3. ESA PSS-01-301 Issue 1, Derating requirements and application rules for electronic components, 1982

4. Electronic Reliability Data: A Guide to Selected Components, Institute of Electrical Engineers Inspec, 1981

5. IEC Standard 68: Basic Environmental Testing Procedures, International Electrotechnical Commission
   IEC 68-1: General, 5 th edition, 1982

   IEC 68-2: Tests

6. CECC 00 107: Quality Assessment Procedures, European Committee for Electrotechnical Standardization
   Part I : 3 rd edition, 1982
   Part II : 1 st edition, 1979
   Part III : 1 st edition, 1980

7. Military Standard 883 C: Test Methods and Procedures for Microelectronics, US Department of Defense, 1983

8. Birolini, A., Möglichkeiten und Grenzen der Qualifikation, Prüfung und Vorbehandlung integrierter Schaltungen, Qualität und Zuverlässigkeit 27, Heft 11, 1982

IMPROVING THE SAFETY OF PROGRAMMABLE ELECTRONIC SYSTEMS

MR. DE KERMOYSAN

The Citroen Car Company

INTRODUCTION

Programmable electronic systems are being used more and more to control industrial production machines.

They allow:

- tight control over complex manufacturing processes
- control over complex movements (robots)
- production modifications to suit changes in the market and in customer requirements
- improved links between production machines and computers.

Without them we would find automation and adaptation very difficult.

But on the other hand their complexity makes it difficult to predict their behaviour when a fault occurs in order to be able to guarantee the safety of the personnel.

At the moment large programmable electronic systems in current usage consist of ten microprocessors and about 50 to 100 kilobytes of memory and they accept information or control the movements of 2000 inputs and outputs. There is now talk of more complex microprocessors consisting of 250,000 transistors and a memory of 1 to 4 megabytes. The reliability of electronic circuits has certainly increased greatly over recent years since the MTBF for certain memory structures has gone from 1.4% per annum to 0.15% per annum but the increase in the complexity of programmable electronic systems is even more rapid than this improvement in reliability.

Under these conditions the concept of safety is no longer the same and we must speak in a totally different language.

A short time ago a prominent member of the French government, the Secretary of State for Prevention of Major Hazards, Mr. Haroun TAZIEFF made a statement on the subject of earthquakes and large natural catastrophes.

"You cannot prevent catastrophes, you have to manage them in advance; too often one sinks back into a false sense of security after the accident, but the people responsible have no right to forget".

Similarly you can also say about safety:

"There is no such thing as absolute safety for a complex electronic system and you cannot prevent a failure but it is vital to manage the consequences of these failures. You must research and verify a relative reliability lower than natural hazards".

In order to make this clearer let's make a comparison between an electronic system and a type of steel.

The steel consists of groups of atoms organised in different regular crystalline structures; this does not stop the crystalline structures having faults and minute fissures in them: but the tensile strength of the steel is as a whole considered to be satisfactory and sufficiently regular to guarantee safety. Similarly, if the crystalline structures of electronic components fail or display anomalies, one could provide at the global level a safe machine, either by limiting the man-machine iteraction or by limiting the effect of a fault to a predictable event known in advance and statistically very unlikely to happen.

We shall describe the following:

- How to improve the relative reliability of programmable electronic systems.

- How to analyse the functional safety of a programmable electronic machine control system and to educe the seriousness of faults involving safety.

RELATIVE RELIABILITY OF A PROGRAMMABLE ELECTRONIC SYSTEM

In order to be able to fulfil its function, even the smallest independent programmable electronic system has the following as shown in Figure 1

- a microprocessor fitted with an internal or external clock

- a set of memories dedicated to:

  operating programs
  user software
  user data

- other variable electronic assemblies.

This system is intended to carry out one function only or a series of functions and to operate effectors by means of different separate systems.

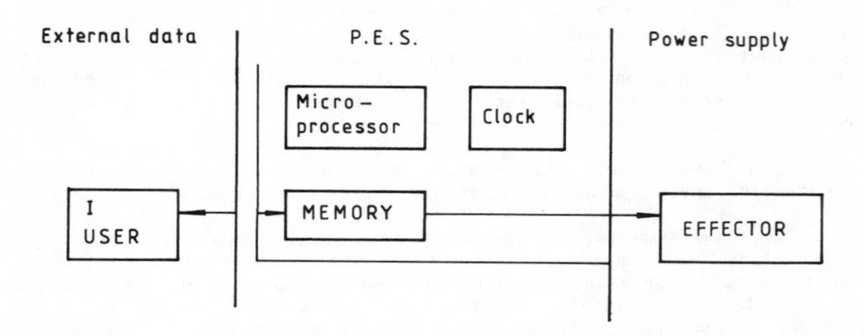

Fig. 1: A Programmable Electronic System (PES)

The programmable electronic system is so complex that it is difficult to calculate its reliability and to guarantee its safety because faults can be very simple.

e.g: the clock stops
microprocessor failure
memory fault
program error

but also very complex and so can effectively modify the way in which the system response to a specific order

e.g: an abnormal link between one group of transistors and another giving a specific abnormal order during a rarely met condition.

Neither visualisation nor a signature check can guard against a fault like this. The signature check does not enable us to check all conditions but indicates that there is for example a 98% chance that the equipment is functioning correctly for the 100 to 1000 accesses envisaged in advance.

Correct operation of the electronic system is ensured by means of a series of regularly performed tests:

- watch dog: activity duration check
- check sum: parity check
- CRC16 check: double check by division
- operation check: bench mark program
- start condition check.

These checks do not always guarantee complete reliability but repeated many times they recognise and warn of behavioural anomalies whether they are caused by a hardware fault, a data corruption or a program anomaly.

PES's known as programmable automatons should have the following checks in accordance with IEC standards:

- a watch dog for the main microprocessor
- check sum on the information in the database
- a start-up condition check during power up.

But these tests are not enough to guarantee worthwhile reliability and real safety in a programmable electronic system: it is necessary to supplement them with all the following tests:

(a) Watch dog redundancy for multi-processors

A check on the current functioning of the watch dog on the main microprocessor by means of a coded phrase sent by one of the other secondary microprocessors to verify the correct functioning of the main microprocessor and vice-versa: this provides redundancy in the watch dog function and true safety.

(b) Start up condition check at regular intervals

Start up conditions are generally verified when the programmable electronic system is powered up. In order to prevent memory loss, electronic equipment cabinets are often provided with a constant power.

(c) Memory check during operation

The standards require a memory check by check-sum at the time of commissioning but certain suppliers offer equipment whose memory verification by longitudinal and transverse check sum is performed during operation at regular intervals.

The designer of a machine will choose equipment of this type for safety because the experiences of INRS have demonstrated that an electromagnetic parasite might quite often modify values in electronic memories.

(d) Protection of series transmissions

All series transmissions, in order to guarantee personnel safety, must include a CRC 16 double check or equivalent which guarantees a higher level of reliability of information transmission (a value of the order of 10^{-8} per annum).

(e) Protection of the information on input-output cards

These cards are generally well protected for slow all-or-nothing inputs by galvanic insulation and by a time-out which eliminates rapid parasites.

On the other hand analogue input cards cannot always have galvanic insulation but it is possible to use differential inputs which limit the effect of rapid parasites and absorb abnormal over voltages.

Rapid all-or-nothing input cards, despite effective galvanic insulation, suffer from the fault of allowing information parasites to enter because the information is at low voltage and absorption of the parasites by the screening on the linking cables is rarely sufficiently effective.

It is necessary to avoid taking into account this information when checking this information and its supplements.

For example the de-coder information must include the values A, $\bar{A}$, B, $\bar{B}$, 0 and $\bar{0}$ and the comparison in the development of these values should be made on the input card before it is taken into account by the main microprocessor. It is vital not to forget to check the pulse number between two passages of the originating mark (0 and $\bar{0}$) to avoid any drift due to a parasite being taken into account in spite of the screening on the linking cable from the coder to the programmable electronic system.

We must remember that according to NICOLAISEN almost 56 robot accidents out of 170 studied are due to uncontrolled robot movements and according to PSA statistics almost 40% of robot shutdowns are due to location information anomalies in the coders.

Therefore in order to guarantee safety of machines involving dynamic positioning by numerical coder, it is necessary:

- either to use originating marks close enough together to prevent important movement anomalies corresponding to a coder data error since the coder information check can only be made on each originating mark

- or to use absolute or semi-absolute coders (Gray code) in which the information corrupted by parasites can only effect the last figure of location information.

(g) Basic software reliability check by injecting uncertain information which the software should detect.

This is the method suggested by Mr. GERARDIN of INRS in NANCY which should be considered as one of the fundamentals for improving software which is becoming more and more important and complex.

We have seen the basic software software for programmable electronic systems greatly increase its detection rate from 40% to 60% then to 70% and then to 80% detection rate of injected errors.

Unfortunately IEC standards have not accepted the idea that basic software must be able to operate but also to detect anomalies introduced into the memory by parasites.

But the user knows nothing about this software which can have faults and cause accidents.

It seems essential to us tht basic software should guarantee 95% detection of anomalies injected into the memory by standardised parasites.

(h) Active redundancy in the safety commands by a programmable electronic system.

Accidents occurring during maintenance or repair operations make up 50% of accidents on automatic machines and 80% of accidents on robots.

Although one can stop the machine for repair it is not possible to stop the machine during maintenance or a robot during maintenance or programming.

In this case the safety of the personnel depends entirely upon the programmable electronic system.

It is necessary to study the safety provided not only by increased reliability but also by redundancy in the safety functions. We shall give an example of redundancy which does not require more equipment but which uses the existing functions which are rarely taken advantage of.

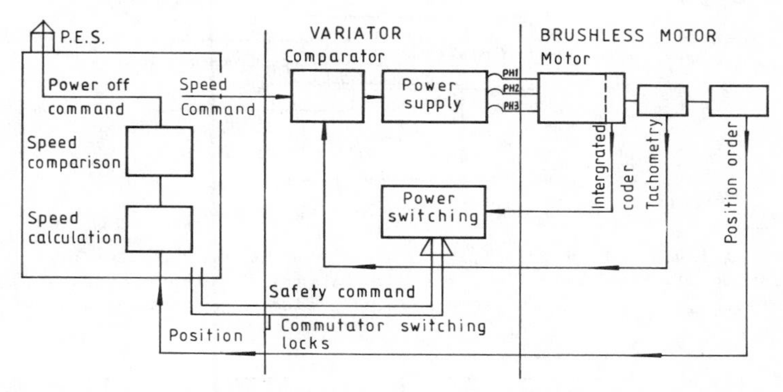

Fig. 2: PES and motor control with safety features

For certain maintenance operations the personnel has to get close to the machine. In this case the programmable electronic system sends a zero speed command which keeps movement to zero by means of servo controls. In order to improve safety it is possible to send via a relay or the PES a power switching lock command (0 phase commutation). This provides redundancy without duplicating equipment.

Finally, if the speed given by the tachometry is different from the speed calculated by the position variation, a comparison of the two speeds allows the program to send a stop command limiting equipment risks. Again this provides redundancy without duplication and without having to increase the amount of equipment or the price.

ANALYSIS OF THE OPERATIONAL SAFETY OF A PROGRAMMABLE ELECTRONIC SYSTEM CONTROLLING A MACHINE OR A ROBOT

The safety of a machine or a robot depends on its components, its design, its construction but also to a large extent on its method of operation.

A machine which operates entirely automatically is not dangerous since there are no personnel near it when it is operating; it is watched from a distance. One can define this with three separate assemblies as in Figure 3:

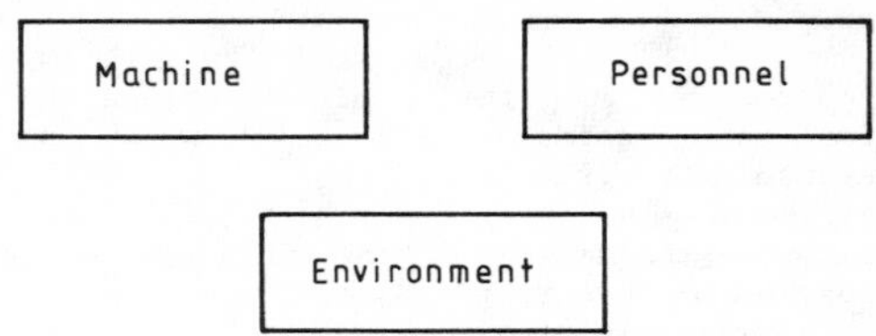

Fig. 3: Three separate assemblies

This is the ideal situation; generally speaking the three assemblies overlap as in Figure 4:

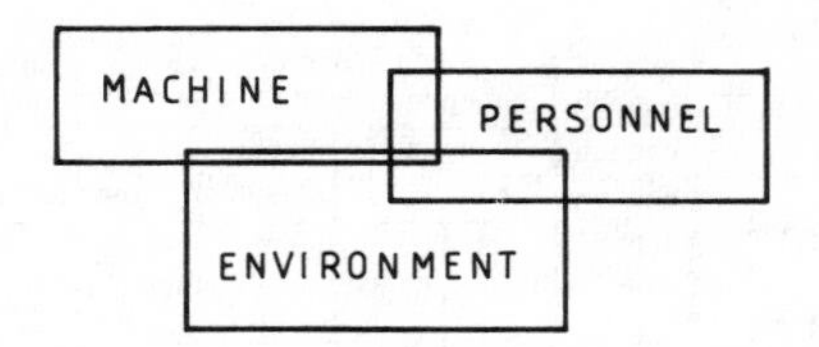

Fig. 4: Overlapping assemblies

A basic safety principle is to prevent the manufacturing process from putting the environment at risk and also the personnel (e.g., explosion, poisoning) by positioning the machine inside a guard as in Figure 5:

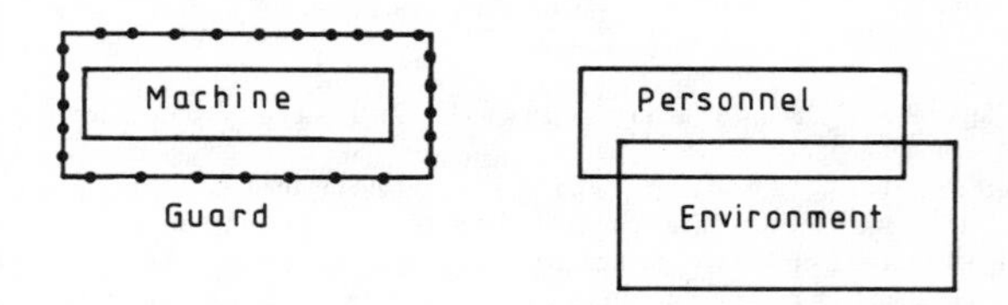

Fig. 5: Guarded machine

Another principle is to concede that an automatic machine must be designed to operate without danger in its normal operating mode: one should separate personnel from the machine by protectors as in Figure 6.

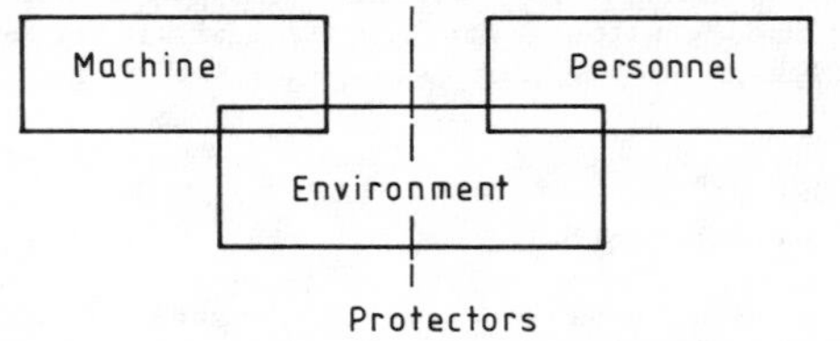

Fig. 6: Personnel separated from machine by protectors

But during assembly, commissioning, maintenance or adjustment it is certain that the guards or protectors will be opened and that personnel will come into contact with the machine and with its hazards.

In order to apply the basic French decree of 10 July 1980 relating to machine safety, it is necessary to study safety in all the following cases:

(a) Normal operation
(b) Operating in each permissible lesser mode
(c) On-going maintenance (lubrication, cleaning)
(d) Programming or teaching
(e) Breakdown and adjustment
(f) Repair (maintenance with dismantling).

The normal operating conditions of automatic machines generally conform to safety requirements by means of the following solutions:

- Case a and b : safety is assured by protectors separating personnel from the machine in a secure manner

- Case b : during a lesser mode of operation clearly set out the work to be carried out by the operator and the specific safety instructions which always require separation of the man from the machine

- Case c : safety is assured by moving the lubrication and cleaning points outside the danger zone or by requiring the machine to be stopped

- Case f : repairs should be carried out with the machine stopped. A protection zone should be provided around equipment which is known to be still in operation.

But during programming, teaching, breakdown and adjustment it is rarely possible to separate the personnel and the machine. it is therefore necessary to study each type of failure and its effects on safety.

The solution lies in applying the principles of quality assurance analysis to the machine from a safety point of view, that is to apply the FMECA method to the safety plan (FAILURE MODES, EFFECTS, CRITICALITY ANALYSIS).

The preliminary safety study quickly points out the dangerous situations and whether they should be eliminated by redesigning the machine or their frequency of occurrence should be reduced by improvement of equipment or environmental conditions: Table 1 is an extract from the reliability values issued by CNET giving details of these improvements.

Therefore if one for example were to replace a programmed electronic system installed in a cabinet without air conditioning fixed to the machine by an air conditioned cabinet fixed to the ground without vibrations and fitted with a parasitic suppressor on the power supply, one would reduce the value of λ by 100.

Table 1: CNET factors for improvement in reliability

Improvement	λ Coefficient of improvement
30° component temperature reduction	5
Supply with parasitic suppressor	5
Elimination of vibrations	5

The gravity of dangerous situations should be studied during the preliminary safety survey from the point of view of safety and of injuries to human beings; and consequently one can determine a weighted coefficient of safety which can be evaluated against a threshold of redesign or improvement of the machine in order to arrive at the desired level of safety.

This method put forward by Mr. LIGERON allows one to "assemble the safety of electronic systems and machines without pre-conditions", by reviewing all possible faults. it can be broken down into several stages as follows:

Stage	Documents	Phase of Work	Results
1	Safety plan	Consideration of study file and maintenance regulations	Analysis of the fundamentals of man machine safety interaction
2	Statement of the problem and safety objective	Definition of the system limits and safety performance	Description of the environmental risks and of the standard operating conditions
3	Preliminary safety study	Preliminary study: Is this system acceptable, if not how can the important points be improved, for example the environmental conditions, a dangerous situation	- Scale of seriousness - Operational block diagram - Check list of dangerous elements - Check list of dangerous situations
4	Qualitative safety study FMECA level 1	Restriction of the system to about 10 to 12 operational blocks in series	Modification of the overall design
	FMECA level 2	Qualitative study of the critical points ignoring the effect of redundancy	Maintenance and programming regulations

5	Lists of dangerous situations	Improvements by means of redundancy	Training
6	Quantitative safety study and quantification	Comparison with the objective (Safety Plan)	Regulations for exchange of worn components

The analysis of the criticality of the safety defects recorded in the FMECA brings in a weighted coefficient Sp between the various safety factors.

This weighted coefficient of safety Sp should represent the risk incurred during operation: the value is defined by the equation: Sp = G + Pa + Pd

where: G = coefficient of seriousness of injuries
Pa = factor related to probability of occurrence (reliability)
Pd = factor related to probability of non-detection

Each of these factors G, Pa and Pd are coded with four weights in accordance with the following criteria:

Table 2: Value of coefficient of seriousness of injuries

Value of G	Seriousness of injuries
1	No effect (negligible)
2	Minor effect without injury (marginal)
3	Medium effect without injury (critical)
4	Serious effect risk of death (catastrophic)

Table 3: Value of factor related to probability of occurrence

Values of Pa	Probability of the failure occurring	Value of equivalent
1	No failure noted after 5 years of operation	$\lambda \leq 10^{-8}$/pa
2	Very few failures noted after 5 years of operation	$10^{-8} \leq \lambda$ 10^{-4}/pa
3	Failure which has caused problems in the past or a complex system not yet fully known	$10^{-4} \leq \lambda \leq 10^{-2}$/pa
4	Frequent failure or new complex system	

Table 4: Value of factor related to probability of non-detection

Value of Pd	Probability of the failure not being detected
1	Very low probability of not detecting the fault in less than 0.3 of a second
2	Low probability of non detection in less than 3 minutes or during one machine cycle
3	Probability of detection only once a day or a 40% chance of non detection
4	Detection difficult and not made automatically

Depending on the value of the coefficient Sp, actions should or should not be undertaken in accordance with the decision table below.

Table 5: Action to be taken

Value of Sp	Results
$Sp \leqslant 6$	Take no account of the failure
$9 \leqslant Sp \leqslant 7$	Provide a daily check or programmed redundancy and establish a safety directive
$Sp \leqslant 9$	Provide redundancy for the operation or a positive fault detector
Sp 10	Review the design

This FMECA study should be carried out for all operational or maintenance conditions defined by the decree of 10 July 1980, that is in each of the cases a, b, c, d, e shown above.

It should therefore form the basis of the study for conformity with the safety regulations which are required of us and should be communicated to the user.

PASS II - PROGRAM FOR ANALYZING SEQUENTIAL CIRCUITS

SCHREIBER, PAUL

Bundesanstalt für Arbeitsschutz (BAU)
(Federal Institute for Occupational Safety)
D-4600 Dortmund 1

ABSTRACT

The ever increasing part of integrated digital circuits used for controlling devices with safety responsibility leads to a growing need for simulation programs suitable for computer-aided failure mode and effects analyses which meet the requirements of safety expertise. PASS II (for process computers) and PASS II-PC (for personal computers) are such special simulation programs. Electric and electronic logical elements considering the propagation time can be simulated with single and multiple faults of the "stuck at" type occurring at any time with due regard of trivalent variables of state (0; 1 and x).

INTRODUCTION

This paper describes a project covering the development of a program system for the failure mode and effects analysis of plant and machine control systems in several steps. The project leader was Mr. Demski of the TÜV Rheinland. The result also takes account of the experiences gathered by Dr. Germer, BIA, and Mr. Six, Technische Universität Braunschweig, when using PASS I in practice.

PASS II (Programm zur Analyse sequentieller Schaltungen) is a generally available auxiliary means for showing, with the aid of a computer, the safety performance of electric and electronic systems equipped with digital logical elements. The FORTRAN program can be obtained as process computer mode (DEC PDP11 series under RSX) or as personal computer (IBM PC AT and DEC Rainbow 100+; each being installed under the UCSD-p-system). More details concerning this type of failure mode and effects analysis and the program can be learned from the research report* and the literature quoted there.

As the computer performance at the work place of the development or test engineer is rapidly improving, this method of showing the safety by means of the failure mode and effects analysis increases in importance.

Simulation programs for logical circuits are usually put to use for the rapid development of complex circuits. These programs can frequently also be used to determine the effects of hardware failures on the control system functioning. Such a failure mode and effects analysis aims at improving the circuit design in order to obtain a greater availability of the plant or machine or to correctly meet the relevant safety requirements in each case, the latter being our object.

PASS II is a simulation program for the latter task only, i.e. for the failure mode and effects analyses. Owing to this specialization it was possible to restrict largely the program scope and the expenditure for teaching the application methods and to prepare the program especially according to the computer-aided safety evidence to be supplied. The increasing proportion of the integrated digital switching circuits in circuits with safety responsibility leads to a growing need for such special simulation programs.

PASS II produces within the computer a software model of the circuit or of a part of this circuit by linking the individual logical elements with one another via the input processor (EPASS). This model is then subjected to a check program, that is the simulation processor (SPASS) computes the

---

* BAU-Forschungsbericht Nr. 406, Adrian, M.; Demski, R.; Kröger, U.; Melchers, W.; Schreiber, P.; Wienandts, W.; Rechenprogramm zur Ausfalleffektanalyse von Maschinensteuerungen, Dortmund 1985

behaviour of the zerodefect circuit and the defective circuits. Both single and multiple fault can be simulated.

COMPARISON OF DIFFERENT METHODS

Before entering into particulars of PASS II I shall discuss in detail the connection with the subject of this symposium, that is the safety of programmable electronic systems.

Germer /1/ compared the following methods for showing the safety of complex electronic controlling devices with one another:

- theoretical fault simulation
- manual fault simulation
- automatic fault simulation
- PASS

Theoretical failure simulation means examining the controlling device from the theoretical point of view. As for the second method the tester manually introduces faults into the controlling device and records the dangerous faults which occur. This method can largely be automated by means of computers and check programs.

The results of this comparison are here presented in an abridged version referring only to PASS II.

PASS II shows the best reproducibility of a test result obtained by different testers. This does not come as a surprise as in this case the image of the circuit formed in the computer is definitely a mathematical model.

The fault covering degree "one", i.e. the number of simulated failures divided by the number of possible faults, cannot be obtained by any of these methods. With the exception of the first one, all methods have a good covering degree. Parasitic effects, however, can only be detected by means of those methods providing for an experimental fault simulation at the original controlling device. What is more PASS II restricts itself to stuck-at faults thus permitting on the other hand the fault model propagation to otherfaults (e.g. disconnection or short circuit; DC-fault), not least because of the possibility to simulate 3 states (0; 1 and X).

Another comparison aspect was the image accuracy, i.e. the conformity of test object and real circuit. The image accuracy of the experimental test methods is good to excellent. For PASS II is was rated good to moderate dependent on the technology.

In our opinion the requirements concerning the testers' qualifications should be very high for all methods, perhaps even somewhat higher for the theoretical ones. As the covering degree for the fault detection is not fully 100 %, a high qualification is requested for that reason alone.

Finally I should like to mention the limits of these methods. PASS II is suitable for digital logical elements only. Analog circuit elements require another method. As a rule that is no great problem because almost in any case several methods are applied during the test in order to assess correctly the complex circuits with safety responsibility. The second limit is the integration degree of the logical elements. In this respect PASS II showed the best results. It is suitable for complicated circuits or circuit parts in relay and contactor technology, for circuits with discrete electronics and for logical elements with small-scale integration up to and including highly integrated circuits of the MSI type as the gate models are often taken for the latter circuits. These logical elements can be combined at will in the circuit.

The higher the "protection degree to be obtained" and the higher the integration degree of the logical elements (up to MSI), the more suitable is PASS II.

It is generally known that none of the investigation methods can be applied to microprocessors and integrated circuits of the LSI type.

According to the model of 5 safety classes recommended by Hölscher and Rader /2/, class 4 corresponds to the "normal safety standard" in technology, this is no dangerous failure may occur in the presence of an undetected dangerous fault. The requirements of the safety classes 1, 2 and 3 are already so high that corresponding circuits can in almost every case only be realised with redundant or partly even diversity hardware und software.

Often the remaining circuit part without microprocessors is so complex that testing with the aid of PASS II would be an advantage. In such a case the data obtained from the microprocessor could be introduced as input data into the software model produced by PASS II. Thus one can analyse which bit combinations at the microprocessor outlet result constantly or only at specific points of time in a dangerous fault.

According to Germer /1/ the tests made hitherto showed that the safety evidence for controlling devices equipped with microprocessors could always be attributed to a lower integration degree because of the redundancy structure.

In this context it should be noted that PASS II permits a comparatively simple testing for multiple faults which is indispensable for circuits of the safety classes 1, 2 and 3.

FAILURE MODE AND EFFECTS ANALYSIS

Assessing the safety of plant and machine controlling devices with the aid of the failure mode and effects analysis will increase in importance in comparison to the other competing or complementing methods; one reason is that this analysis can very effectively be carried out in formalized form with computers and Boolean algorithms, thus leading to reproducible results.

Unlike the fault tree analysis the failure mode and effects analysis detects systematically all effects - including those considered to be harmless - which can appear for all possible failure modes of a logical element and that successively for all logical elements.

Hence the result of a failure mode and effects analysis is the description of each element behaviour in case of a failure. If the element under consideration is part of a plant, the result of the failure mode and effects analysis of this element (e.g. an electronic componentry) is a catalogue of its possible failure modes which is then used for the failure mode and effects analysis of a larger unit (e.g. a plant component). When all these analyses have been completed in hierarchical order, from the smallest to the largest unit, the whole failure mode and effects analysis describes the plant failure behaviour, which can be compared with a stipulated behaviour list. The failure behaviour having thus been completely recorded, one can assess the hazard emanating from all failure modes. Supposing a complete safety evidence is requested, it will usually be furnished by means of a failure mode and effects analysis in spite of the greater expenditure. A fault tree analysis, however, cannot show the completeness without having recourse to a failure mode and effects analysis.

PASS

The target groups for PASS II are therefore those institutions which have to give safety expertises on circuits, but also the research departments of enterprises which design plant and machine controlling devices and often already make use of process computers and efficient personal computers.

There are consequently a few important marginal conditions which have to be met in order to use generally available computers (minimum word length 16 Bits):

- highly portable software
- limitation of the number of circuits to be simulated (nominal machine and faulty machines) to 200
- limitation of the number of circuit inputs (test patterns) and outputs to a total of 64
- limitation of the number of logical elements to 200
- simulation of gate running times without consideration of the transitions.

The above-mentioned limits are not set by the program, but by the computer (main memory, program run times) and can rather easily be expanded.

To offer a few rough clues for such program run times:
The analysis of a simple network takes about 1 minute. A network with 40 logical elements requires about 3 minutes and a "demanding" simulation 15 to 30 minutes.

Because of the use of trivalent variables of state (0; 1 and X) the fault type "stuck-at-X" could be added to the classical stuck-at fault model. Thus it is possible to simulate circuit conditions which result from undefined input conditions (e.g. when switching on the machine), from problematic part circuits with inadmissible input bit combinations (e.g. RS-Flip-Flops) or - a last example - from logical element failure modes having no clear failure direction.

Furthermore faults of the "INVERS" type (should-be signal inversion) can be simulated.

Figure 1 Truth table for trivalent states

Bit 1	Bit 2	variable of state
0	0	X
1	0	not used
0	1	0
1	1	1

Figure 1 shows the greater expenditure for the registration of the 3rd state (2 Bits instead of 1 Bit). This expenditure propagates, e.g. in the logical equations for the logical elements and in the algorithms which are necessary for checking and determining the gate run times.

On the other hand the introduction of the undefined state X excludes possible random results which could perhaps lead to a dangerous wrong assessment of a circuit. Taking account of the state X when simulating circuits with safety responsibility is therefore an important precondition for a correct safety evidence assessment.

Figure 2 Parallel fault simulation of trivalent variables of state

Operand m

0	1	2	3	. . . .	12	13	14	15	Bit
1	0	0	1	. . . .	0	0	0	1	word 1
1	1	0	1	. . . .	1	0	0	1	word 2
1	0	X	1	. . . .	0	X	X	1	variable of state
0	1	2	3	. . . .	12	13	14	15	circuit
-	1	2	3	. . . .	12	13	14	15	fault

Figure 2 shows how the 16 Bits of a data word can be used in the parallel fault simulation technique to simulate simultaneously - in this case for the operand m - both the faultless circuit (nominal machine, represented by the circuit 0) and 15 other faulty machines which results in a correspondingly shorter computing time.

INPUT PROCESSOR EPASS

The program system is composed of 3 independant processors.

Via the input processor the network, the Bit pattern combinations (test patterns) at the circuit inlet and various definitions for simulation and fault production are fed into the program.

Figure 3 Network element

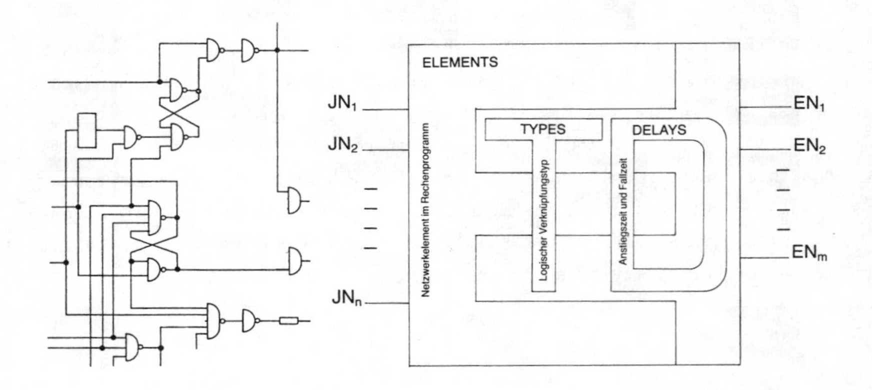

Figure 3 shows a network element (circuit element). It ist characterized by the logical type of the circuit element and a characteristic delay time (rise time and fall time) which depends on the technology and the circuit element type. The circuit structure is set up by the input of the circuit elements interconnections.

The most important circuit element types (e.g. gates, shift registers and Flip-Flops) are listed in a logical element library. Other circuit element types can be assembled. When allocating the delay time one makes use of the fact that only a few digital logical elements differ from one another with regard to their characteristic time behaviour.

Faults for the simulation of a faulty circuit (faulty machine) can be introduced as automatic fault generation or by presetting each single or multiple fault individually. The failure occurrence times can be chosen at will, that is general all-including and special simulation runs are possible.

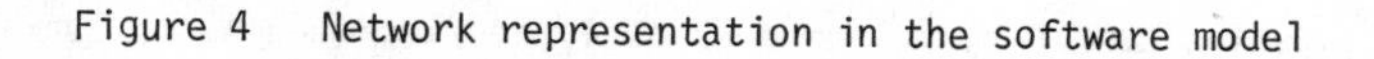
Figure 4 Network representation in the software model

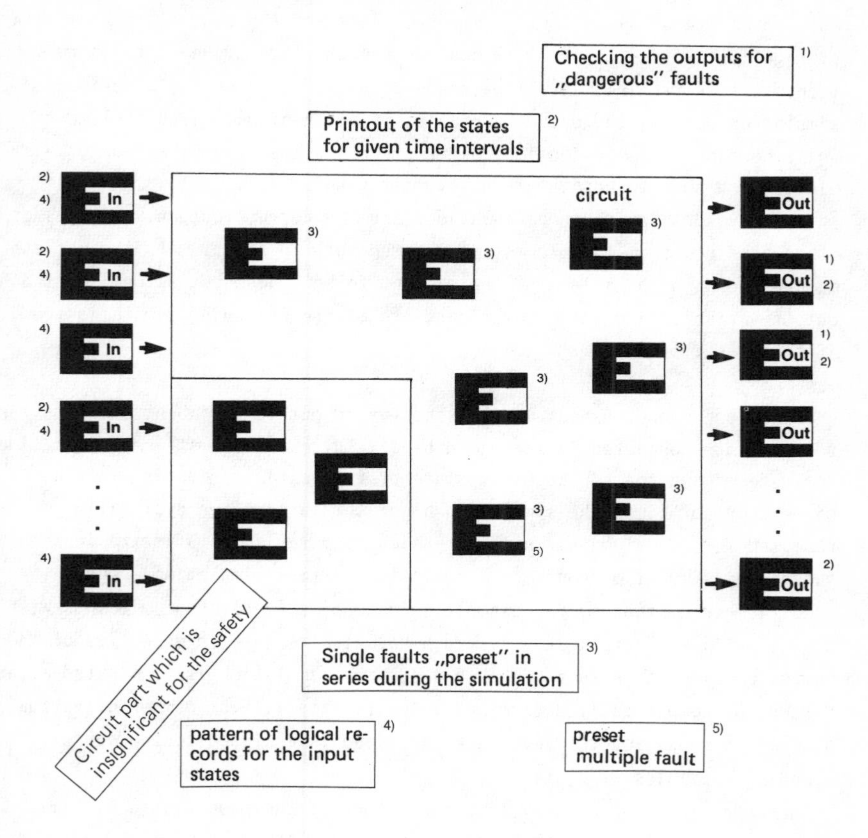

Figure 4 explains the above-mentioned connections. The elements "In" and "Out" characterize special circuit elements in the software model which can easily be logically connected with other elements.

SIMULATION PROCESSOR SPASS

All these data (i.e. the data of the nominal machine and of the faulty machines) are processed by the simulation processor SPASS.

The simulation procedure is based on the fundamental assumption that the network condition can completely and uniquely be described by the variables of state of all circuit element outlets and the contents of the delay storage (delay time counter) allocated to each element. Hence the simulation has to determine, at all discrete points of time during a defined simulation period, the output variables of state of all circuit elements while considering also the logical function and the time behaviour of each element and then to update the delay time counter.

Input values for these calculations are the current output variables of state obtained from the elements connected both in series with the outlets before and with the inlets of the element to be processed, as well as the output variables and the counter contents of the preceding simulation run, if necessary.

Since one cannot expect in all cases valid outlet variables of state for the elements connected in series to be already available at a specific time for determining the variables of state of a specific circuit element at a given point of time, the simulation algorithm is so often cyclically repeated for all elements until all calculated variables of state in two successive simulation runs are identical (virtual simulation). Under specific circumstances, for example in case of a feedback of elements with the delay times "zero", several truth values alternate irrespective of the number of simulation runs. In such cases the simulation is terminated after N runs (N = empirically determined value for the maximum number of virtual simulation runs) and the value X = undefined is assigned to the corresponding variables of state.

Here we cannot go closely into the matter of the problematic of time intervals determination, that is the question of how to define the discrete points of time for a new calculation of the circuit states. Account has to be taken of components with different high speeds and occasionally even with rather slow speeds. Intermediate points of time are for example necessary to record completely the possible effects of single and multiple faults during the fault simulation.

Via the output processor APASS the simulation parameters and results are placed at the user's disposal. The output functions are tailored to the tasks which have to be carried out for proving the safety of circuits. One of the most important output functions is the list of all detected faults with or without mask. Such search masks can be used to establish and to print separately characteristic output states, e.g. dangerous faults.

PASS-II-PC PROGRAM

Having dealt so far with the application possibilities and properties of the program for analyzing sequential circuits I would now like to consider the PASS-II-PC program.

As there are many special questions which arise only during the actual test, a free availability of a time-unlimited calculating capacity at the work place is desirable. Efficient personal computers offer this capacity even for comprehensive programs (PASS-II-PC ≧ 400 KB backing storage demand; directly adressable storage ≧ 256 KB), the nonrecurring costs being comparatively low.

Owing to the great type diversity of the personal computers this program should have a high portability. An accurate program maintenance ist necessary for all programs concerning safety questions. If possible, only one version should be maintained that way. This rules out many parallel program versions.

About 10 years ago the UCSD-p-system was developped at the University of California, San Diego. This system is presently available for approximatively 150 computer types and permits the generation of highly portable software.

This feature is obtained with the aid of a special program object code, the p-(Pseudo-)Code. When compiling a program written in UCSD-FORTRAN 77, this p-Code is generated instead of the specific computer code. It is then translated into the absolute code during the run time on the computer at hand by means of computer-specitic interpreters.

PASS-II-PC with the p-Code was installed on the computer types DEC Rainbow 100 + and IBM PC AT 02.

This was a brief comment on the analysis program PASS II which could be an important aid for those persons giving expertises on or developping electronic safety circuits. However, other methods should also be applied at the same time in order to record the consequences of parasitic effects and to be able to obtain a better assessment of the simulation accuracy of the software model.

REFERENCES

1. Germer, J., Vergleich verschiedener Methoden zum Sicherheitsnachweis von komplexen elektronischen Steuerungen, Proceedings of 2nd Colloquium of the Section for Research of the ISSA, SAFETY AT WORK WITH AUTOMATION AND NEW TECHNOLOGIES, Bonn - Hennef/Sieg 8-10 Mai 1985, p. 220

2. Hölscher, H., Rader, J., Mikrocomputer in der Sicherheitstechnik, Verlag TÜV Rheinland, 1984

EXPERIENCE WITH COMPUTER ASSESSMENT

G.Glöe and G.Rabe
Technischer Überwachungsverein Norddeutschland e.V.
Grosse Bahnstrasse 31
D-2000 Hamburg 54
Federal Republic of Germany

ABSTRACT

Ever since the beginning of the seventies we have been examining computers with the aim to prove the safety, to increase their reliability and, if necessary, to provide prerequisites for licensing. An overview of these systems is given. The qualification procedure for computers in Germany and over ten years of experience with the use of such procedures are outlined. Two research projects, the one involving a safety analysis of computer software and the other concerning the development of software safety tools, are described in detail. Using a pipeline control system as an example the procedures for the safety proof will be shown.

OVERVIEW OF COMPUTER AIDED, SAFETY RELEVANT SYSTEMS

In german commercial light water reactors, computer aided systems are employed for tasks with varying degrees of safety relevance. Comparatively small microprocessor systems, using programs requiring only a few kilobytes of memory, are used for certain tasks of the reactor protection system. Similarly sized microprocessor systems are used in the post-disturbance instrumentation to determine the departure from nucleate boiling.

Larger microprocessor systems, with programs requiring several tens of kilobytes of memory, are used in the radiation protection system.

Full scale process controlling computers are used for on-line, closed loop control of the motion of the control rods. These computers will be described in more detail in connection with the experience with computer qualification. In-service inspections of parts of the reactor protection system are also performed with the aid of process controlling computers. Finally, extensive process controlling computer systems are used for operational and disturbance recording in order to facilitate the reconstruction of disturbances.

With the increasing speed of computer systems, coupled with their decreasing cost, computer aided systems are penetrating ever deeper into application areas that go beyond the scope of these examples. Pipeline control systems are mentioned here.

COMPUTER ASSESSMENT IN THE LICENSING PROCEDURE

The licensing procedure in Germany requires that components used in nuclear power plants shall be qualified before they are employed. A whole set of rules for general aspects of technical computers has existed for some time. However, the safety aspects that are so eminently important in nuclear power are not specifically treated in these national rules. Work has been begun on a standard "Principles for computers in systems with safety responsibility". These principles shall be applicable to computer systems in all safety relevant applications, not only nuclear power.

Due to the lack of binding rules and regulations that can be referred to for computers, the qualification of computer based systems is done according to the state of the art in the following steps:

The requirements for the computer system must be defined. They can be checked with respect to completeness and consistency. For the subsequent steps, the requirements' definition must also be clear, sufficiently detailed and unique; these properties are also checked for.

After the requirements have been defined, the system design can be examined. It must, of course, correspond with the requirements' definition, be free of contradictions and be complete. It should also be examinable.

The hardware is examined for compliance with the principle demands made on hardware for safety relevant tasks concerning its reliability, its time behaviour and its behaviour under external influences. Within the framework of prototype and aptitude tests the proof that particular components fulfil the safety requirements can be given. The manufacturer can then produce exactly this kind of hardware and the user can apply it without further testing.
Beside these tests it is customary to perform functional tests of the actual hardware components and the whole system using them.

When qualifying computers, one must not only examine the hardware, but also the software! There is a number of methods available for furnishing the proof that software has the required properties. They include program analysis, statistical examinations, white- and black-box testing etc. Which method is to be applied depends on the desired depth of the examination, which in turn depends on the safety relevance of the application.

Isolated examinations of the hard and software by themselves do not necessarily imply that the integral system will also have all the required properties. For this reason the qualification procedure for computers also includes a test of the whole system under real-time conditions.

Finally the procedures for maintenance and modification of the computer system must be determined. This guarantees that a system that has been deemed safe will remain in that state despite alterations that experience may necessitate.

EXPERIENCE WITH COMPUTER ASSESSMENT

The Technischer Überwachungsverein Norddeutschland (Technical Supervisory Association of Northern Germany - "TÜV") began working on computer qualification in 1971 with the certification of a reactor protection computer. At that time, the required reliability of those computers could not be demonstrated, so that they were not employed on-line. However, in order to get an impression of how powerful the reactor protection computers really were, they were run on-line open loop, parallel to the conventional reactor protection system, for a number of years in two power plants. The evaluation of the operational experience, which was done by the Gesellschaft für Reaktorsicherheit (Company for reactor safety - "GRS"), revealed no conspicuous behaviour that would have spoken against using them.

The knowledge gained there was utilised two years later by the assessors as well as the manufacturers during the realisation of the control rod steering computers. These computers have now been in use in several german boiling water reactors for many years. Because it is not a redundant system, the computer performs several self-tests and shuts itself off if it detects errors (having switched rod steering to manual control first). Results of reliability evaluation are described in /1/.

Ever since radioactive steam escaped from a german nuclear power plant in 1978 and the documentation system was down, the reliability and availability of such systems has been given more attention. The operation and disturbance recording is now done with instrumentation containing an extensive process controlling computer system. We have evaluated the data produced by this system a number of times. It became clearly evident that such systems are of little value if they do not include powerful evaluation tools. Such tools would however also give an insight into the reliability of such data collecting computers: the reliability of some machines is noticeably below the level of the comparatively small machines dealt with here.

Since 1982, microprocessor systems have also been or are being qualified for use in nuclear applications, starting with applications in the radiation protection system. The problems that have arisen with computer systems have not been more serious than the problems that otherwise would be regarded as "normal" and, based on our present experience, we do not expect unusual difficulties in the future.

The computer systems described so far were all intended for use in nuclear power plants. The safety requirements are particularly high there, but the basic principles applied to computer qualification for nuclear applications can, of course, be used in other areas too; a lower degree of safety relevance will result in a less rigorous qualification procedure.

The purpose of the pipeline control system is to control and monitor the transportation of crude oil through two pipelines. For safety and availability reasons a redundant process controlling computer system is used. An important component of the pipeline control system is a microprocessor controlled remote action system which transmits all measured values and commands.

The examination of the pipeline control system was done in two parts: a purely computer specific examination and a systems engineering examination. The computer examination aimed to show that the hardware and particular parts of the software fulfilled the requirements that were made not only for safety, but also for ergonomic reasons. All these examinations were practical tests performed directly on the computer system (black box tests).

In the systems engineering tests it became evident that it was possible to simulate all the permissible and forbidden operational conditions for testing purposes using the principle process variables pressure, flux and temperature. For this reason we chose a black box test to demonstrate the correct behaviour of the system under forbidden conditions.

For all the functions input quantities were defined resp. altered as is customary in black box testing. It was checked to see if the system performs according to its specification. When errors were encountered in this phase the test sometimes reverted to a desk top analysis.

AMOUNT OF EFFORT

The effort involved in examining a redundant, very safety relevant computer system was about 10 man-years in the early seventies. The programs of this system, which were written in assembler, and the data comprised together approximately 12 kilobytes.

Meanwhile, progress in the fields of computer examining techniques - in particular in software engineering and examination techniques - has led to a significant reduction in the amount of effort. For safety relevant systems the effort for a 20 kbyte assembler program is today about 4 man-years.

The currently proceeding examination of a system with a low

safety relevance that is written in a higher language requires about 15% of the development effort. The effort for the program analysis lies between 50% and 100% of the effort for program development.

RECOMMENDATIONS

Based on our experience with computer assessment, we feel that adhering to the following recommendations is of particular importance for a rational and efficient qualification:

- The use of operating systems in computer applications is to be recommended due to their observed low failure rates. This applies only when quantifiable operational experience is available.

  Since the use of an operating system can relieve the application programs of administrative tasks their scope will be considerably reduced. Thus, this recommendation should lead to a reduction of effort and an increase in safety.

- A formalised requirement's specification should be made for computers with safety relevant tasks.

- The processing speed and memory capacity of computers should be chosen such that the programming solution of a task only depends on the task and not on the available resources.

- Commands that perform variable operations or lead to variable continuation points should not be used.

- Computer architectures should be used that make a dynamical alteration of the programs impossible. They should permit reading the invariant program- and data-areas for testing purposes.

- Interrupts should only be used when the requirements concerning reaction speed cannot be met otherwise.

RESEARCH AND DEVELOPMENT WORK ON SOFTWARE VERIFICATION

The control rod steering computers already mentioned were subjected to a qualification procedure that was not as rigorous as it would have been for a reactor protection system. This was due to the lower safety relevance of the control rod steering system. However, due to its operational success, it was chosen as the object for a rigorous post-qualification procedure in order to determine the effectiveness of the qualification methods.

The qualification procedure was sponsored by the German Federal Ministry of the Interior as project SR 293 and conducted as a cooperation between TÜV, the GRS and the power plant company Kraftwerk Union (KWU). The aim was to determine:

- Which methods and procedures are practicable to prove the reliability of software?
- How successfully can this proof be furnished?
- How great is the effort involved?
- How can software qualification be made easier?

During this project, which was concluded in 1984, we realised that large portions of the program analysis procedure could be automated. In order to achieve this, we applied for funding to develop appropriate tools together with the GRS and the OECD Halden Reactor Project in Norway. This project is called Software Safety Tools, abbreviated SOSAT, and the contributions of TÜV and GRS are sponsored as the German Federal Ministry of Research and Technology's project number 1500 679.

In analysing safety relevant programs we must start with the machine code, because this is what is actually running on the computer. In order to make the analysis tools universally applicable, it is necessary to transform the machine codes of different processors into a uniform notation that is processor independent. For this notation we have chosen a general assembly language whose syntax and semantics have been defined within the SOSAT project. Following this definition, we have begun work on a universal, intelligent disassembler: universal, in that it will be able to recognise and translate different machine codes, and intelligent, in that it will automatically be able to differentiate between code and data without having to be manually restarted when it encounters "untranslatable" machine code.

As a check that the disassembler has worked correctly, the inverse transformation is intended, at least for an example. With the exception of data areas the original and the reassembled codes should be identical. We will then have the possibility of transforming the machine codes of any computer into a common assembly language. All analysis tools based on this common assembly language will then be able to be applied.

Simultaneously, work is being done to adapt the PEARL Analyser developed by GRS to this assembly language. The PEARL Analyser presently analyses PEARL programs (PEARL is a rival to ADA!) and determines their control flow structure, within certain limits the data flow and produces information for generating data to test all paths through the program. It is intended to adapt other analysis programs as far as possible or available!

CONCLUSION

The practical experience gained during the examinations we have performed has revealed that the safety proof for computer systems with safety relevant requirements can be given. The procedures we have described make this proof possible.

The greater the safety relevance of a system, the greater must be the depth of examination. The total amount of effort and expense increase with the examination depth. Thus economic aspects become increasingly important as the safety relevance of the system increases. The economical aspects may not predominate over the safety requirements.

REFERENCES

/1/ W.Ehrenberger, J.Märtz, G.Glöe, E.-U.Mainka
Reliability Evaluation of a Safety Related Operating System
SAFECOMP 1985 Proceedings

SAFETY ASSESSMENT METHODS FOR NEW AGR FUEL ROUTE CONTROL SYSTEMS

A BRADLEY

National Nuclear Corporation Limited
Booths Hall
Chelford Road
Knutsford, Cheshire
England

ABSTRACT

Extensive use is made of programmable electronic systems to provide sequencing and safety interlocking of fuel handling plant on the new Advanced Gas-Cooled Nuclear Reactors currently under construction at Heysham (England) and Torness (Scotland). Adherence to numerically based safety criteria must be demonstrated for these systems. Following a brief description of the fuel route through the station this paper gives a more detailed description of one particular system (the fuelling machine control system) and outlines the general approach to and methods used in a safety assessment of that system. These methods cover hazard definition, risk allocation, treatment of random and common cause hardware failures and software reliability.

INTRODUCTION

The twin Advanced Gas-Cooled Reactors (AGRs) at Heysham and Torness, due to be completed in 1987, are the most modern in a series of more than 40 power reactors constructed in the UK. The National Nuclear Corporation Limited (NNC) and its predecessors have been involved in building the majority of these reactors and were responsible in most cases for the design, procurement, construction and commissioning of the complete power station.

In contrast with most other designs of nuclear reactors outside the UK, the AGR provides for on-load refuelling - i.e. replacement of fuel assemblies whilst the reactor is operating at power. Withdrawal of a used fuel assembly and insertion of a new assembly at the reactor is performed by the fuelling machine, a mobile, shielded pressure vessel containing a hoist system and fuel storage tubes. The atmosphere within the reactor, and also within the fuelling machine when connected to the reactor, is CO_2 pressurised to 40 bar a. This atmosphere is also maintained within the irradiated fuel decay store, (Fig 1) where ex-reactor fuel assemblies are kept and cooled for 28 days before being transported by the fuelling machine to the irradiated fuel dismantling cell. At this facility the fuelling machine is blown down to atmospheric pressure before the assembly is lowered into the cell for dismantling. During the dismantling process individual fuel elements (8 per assembly) are despatched to the fuel storage pond where they are stored under water in skips for a further 80 days before the skips are sealed in flasks for transport to the fuel reprocessing plant. The fuelling machine transfers the re-usable upper section of the fuel assembly from the dismantling cell to a maintenance cell for refurbishment. The station refuelling schedule requires that such complete refuelling cycles are completed at a rate of approximately 2 per week.

Within the fuel route there are 4 main centres of plant control and interlocking, namely the fuelling machine (the control system is carried "on board"), the irradiated fuel dismantling cell, the ponds and the maintenance cells. Because of its varied duties the fuelling machine control system is the largest and most complex of those listed and is considered in detail in this paper. The same philosophy and methods, however, have been applied to all 4 systems.

The fuelling machine itself is a large item of plant, some 30m high and weighing 1200 tonnes. Since it forms part of the reactor pressure boundary, when connected, and hoists an assembly of highly active fuel through 30m, interlocks and protective features of high reliability must be provided in order to ensure that fuel is not damaged in coincidence with a breach of the pressure boundary. A range of interlocks of intermediate and lower reliability are also required and these safety requirements are in addition to comprehensive functional requirements dictated by routine sequence control and indication. The scope of these requirements is reflected in a simple I/O count on the control system which shows approximately 3000 digital inputs and approximately 500 drive and indication outputs.

Figure 1 General Arrangement of Fuel Route

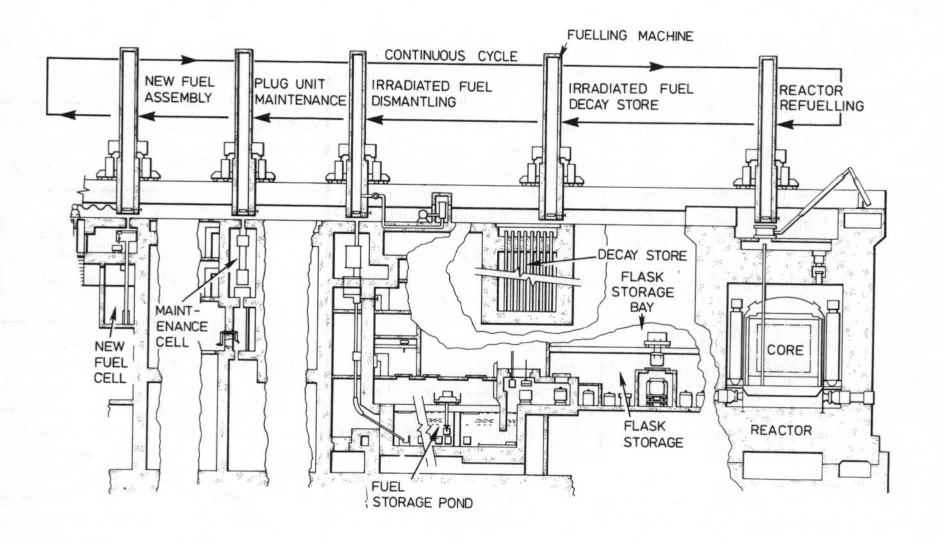

SAFETY REQUIREMENTS

The principle safety design requirements of a nuclear power station relate to the potential risk of a radiation hazard to the general public. For the latest UK reactors quantified targets are formulated by the Generating Boards (1) in order to help in ensuring that any risks from nuclear plant compare favourably with other risks from everyday activities. In particular, the Central Electricity Generating Board (CEGB) state inter alia that

(i) For any single accident which could give rise to a large, uncontrolled release of radioactivity to the environment resulting from some or all of the protective systems and barriers being breached or failed, then the overall design should ensure that the accident frequency is less than 10^{-7} per reactor year.

(ii) The total frequency of all accidents leading to uncontrolled releases, as in (i) above, should be less than 10^{-6} per reactor year.

In practice the second target, covering as it does the whole station, is more restrictive and leads, on the fuel handling route, to a target effective frequency of approximately 10^{-8} per reactor year, applied to individual fault sequences which could lead to an uncontrolled release of radioactivity.

In addition to the safety design requirements relating to uncontrolled releases of radioactivity, the CEGB has issued guidance for fuel route interlock and control systems, in particular defining and classifying interlock integrities required for fuel route plant in terms of the potential hazardous consequences of interlock failure. Table 1 summarises our interpretation of this guidance indicating safety classes (A,B,C), hazard definitions and target annual frequencies for each degree of hazard.

Table 1 Fuel route hazard classifications

	Class A Frequency $<10^{-6}$/year	Class B Frequency $<10^{-4}$/year	Class C Frequency $<10^{-2}$/year
Off-site release of radioactivity	Uncontrolled *	Intermediate	Low
Radiation dose to operators	Serious	Above annual limit	Within annual limit
Coolant (CO_2) release	Extreme temperatures and toxicities	Intermediate but cannot be isolated	Can be isolated
Plant damage	Permanent, major plant shutdown	Plant shutdown for a long period	Plant shutdown for a short period

* Designated A+ hazards, subject to overriding frequency requirements discussed in text

The structure provided by this classification scheme, together with a detailed knowledge of the plant and plant operations, is used to produce a hazard schedule for each of the four major areas of control and interlocking. Each hazard in the schedule is a fault sequence (usually associated with individual or small groups of plant drives/actuators) whose potential consequences are classified according to Table 1. The frequency requirements of Table 1 are placed on these individual fault sequences. The completeness of the hazard schedules relies greatly on experience accumulated on similar plant over many years of reactor operation.

PRELIMINARY HAZARD ANALYSIS

In order to derive the required reliabilities of electrical interlocks a preliminary analysis is performed and is based upon the hazard schedule described above. This analysis can be summarised as the crystallisation of required electrical interlock reliabilities taking account of demand rates on the interlocks and any mitigating effects or mechanical interlocks that would reduce the probability of the hazard occurring should the electrical interlocks fail.

A simple formula is applied to each hazard to derive the interlock failure probability P_I

$$F = D\ P_I P_M \qquad [1]$$

where F is the hazard frequency requirement, D the demand rate on the interlock and P_M the probability of hazard occurrence should the interlock fail (i.e. mitigating effects/mechanical interlocks). It should be noted that when the demand rate on the protection system is high in relation to the frequency of system proof testing, the product of the demand rate and the interlock failure probability must be replaced by the interlock failure frequency - i.e. the required interlock reliability then takes the form of a failure frequency.

Figure 2 Preliminary Hazard Analysis Schematic

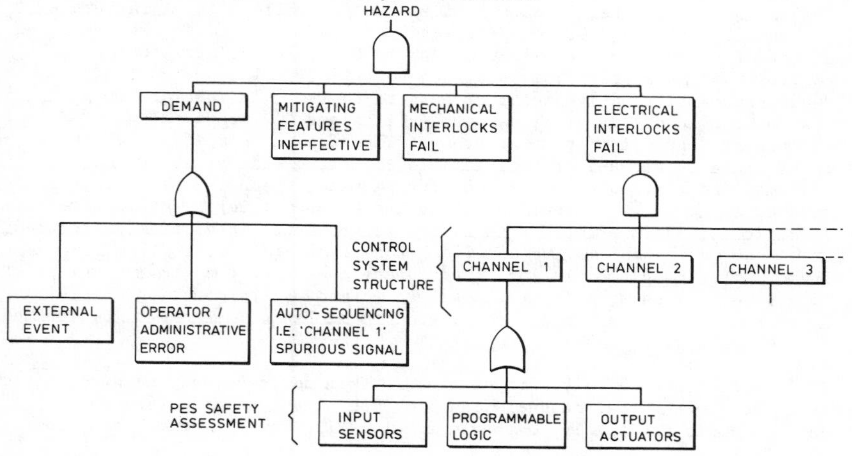

Fig 2 represents the decomposition of the hazard frequency according to equation 1 and it can be seen that three sources of "demand" are identified. Firstly external events whose consequences are to be limited by interlocks - e.g. the occurrence of a snag during hoisting of a fuel assembly out of the reactor. Such demand rates are estimated, wherever possible, from operational experience accrued on similar plant. Secondly, operator errors and/or failures of administrative controls/procedures resulting in requests for actions which may be unsafe - e.g. an attempt to place a radioactive fuel assembly into an unshielded facility. Thirdly, spurious signals from the automatic control equipment which would cause plant drives or actuators to operate out of sequence if not checked. In multi-channel systems a primary channel is designated the control channel and its spurious signal rate is assessed. Channels 2, 3 ... are designated interlock channels for this type of demand although account must be taken of potential causes of common mode failures (CMF) which could both initiate the demand in the primary channel and disable interlocking channels.

Mitigating features are classified as aspects of mechanical or electrical plant which are not engineered as interlocks but are inherent (e.g. electric motor torque limits, component strengths, etc) whereas mechanical interlocks are viewed as additional plant introduced to provide independent protection against initiating faults (e.g. slipping clutches, automatic locks, etc).

The required electrical interlock reliabilities comprise the interlock schedule from which the coarse control system structure is derived (Fig 2). Those channels based upon programmable logic, including associated input sensing devices and output actuators are considered in detail in the following sections.

CONTROL SYSTEM STRUCTURE

Based upon experience of "conventional" protection system reliabilities and presupposing, to some extent, that microprocessor based interlocks can be engineered to be at least as reliable as electromechanical interlocks, simple channel counting, allocating a failure probability on demand of $\sim 10^{-2}$ to each channel, is used to determine the number of channels and their interlock content. On all the fuel route control systems channel 1 is designed to satisfy all functional requirements (i.e. drive/actuator outputs control, indications outputs, push-button inputs) as well as low level (class C) interlocking by monitoring inputs from plant sensors. The size and complexity of these tasks dictates the use of programmable logic for channel 1. On most of the systems, and particularly the fuelling machine, channel 2 is similar in structure and provides duplication of the channel 1 logic for intermediate level (class B) interlocks - the software, however, is diversely produced from that on channel 1. CMF considerations limit the reliability achievable from further programmable logic redundancy and hence the limited number of high level (class A, A+) interlocks are formed by the addition one (class A) or two (class A+) "hard-wired", electro-mechanical channels, designated channels 3 and 4.

Each channel receives input signals from devices independent of those associated with other channels and provides output signals through devices dedicated solely to that channel. Final output contactors from each channel are arranged in series so that all must give a permissive signal before an operation can commence. Furthermore, should any one line indicate a hazardous condition during normal operation, the drive is stopped. A general design principle has been applied in that failure to energise an output or failure of an output whilst energised is a safe failure - for example hoist brakes are applied when de-energised.

Figure 3 depicts the coarse structure of the fuelling machine control system and highlights a number of additional safety related features. In particular all final drive contactors on channels 1 and 2 are fitted with auxiliary contacts which feed back as inputs to the logic and are continuously scanned and compared with internal logic stages. Also, each plant sensing device provides two digital inputs to its channel, one normally open and one normally closed, and each pair is regularly checked for complementarity. The Monitoring Computer is a third microprocessor system arranged to monitor and receive information from the two main channels, displaying the current state of software sequences on a VDU and, when necessary, error and diagnostic messages on a line printer.

Figure 3 Fuelling Machine Control System Structure

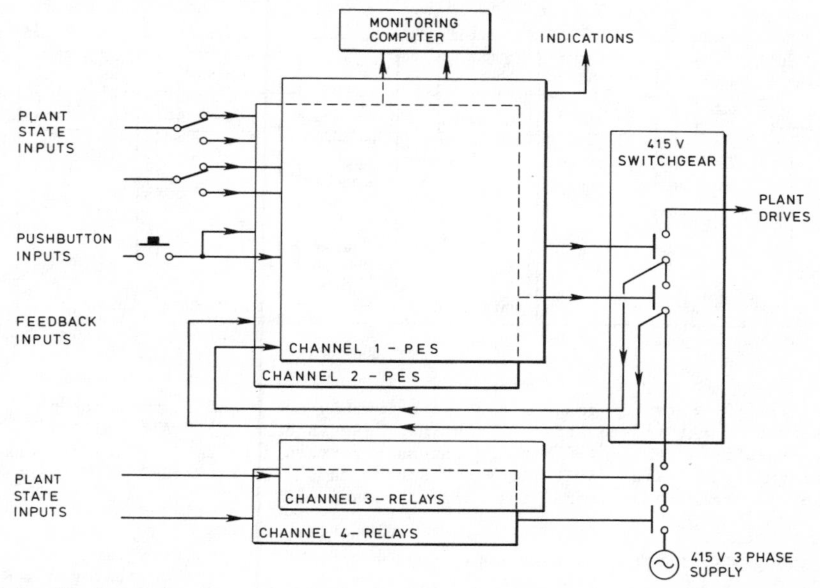

This third system also acts as an input device for manual overrides which can be applied to inputs on either of channels 1 and 2. This feature, used only under strict administrative control involving key interlocks, is designed to permit recovery from fault sequences which have terminated safely and after which the plant must be brought to a state in which repairs or replacements can be made. During normal operation the monitoring computer functions in a passive role having no effect on the control and integrity of the fuelling machine and hence is not included in the reliability analysis.

The simplified internal structure of one of the programmable channels is shown schematically in Figure 4. All inputs are terminated in groups of 16 on modules which are themselves arranged in groups of up to 16 on racks. The ten input racks on the fuelling machine channel 1 are all connected to the main rack which contains the main processing unit in association with the sequence software in EPROM. Further connections exist between the main rack and four output racks from which emanate the sequencer outputs.

Each input and output module contains its own CPU and 24K-bytes of memory. Each input feeds via an optical isolator into a data multiplexer and the input cards carry out regular self tests by transiently forcing the optical isolator inputs all high or low so that the processor can check their operation without affecting the plant. Output modules contain 16 normally open contact reed relays and each has an LED showing energisation of the driver coil. The status of these is fed back and the coils regularly pulsed as part of the module self diagnostics.

Figure 4 Fuelling Machine Channel 1 Schematic

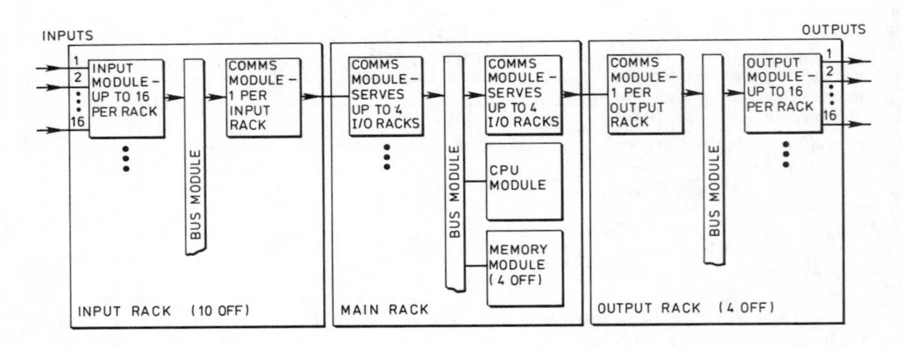

All racks contain independent 5V power supply units and communication between racks is via serial lines (and is asynchronous). All modules containing a CPU employ a relay watchdog refreshed by the on-card software. Diagnosed, on-line faults from any source cause all I/O modules to be reset to a known safe condition (all outputs low).

HARDWARE RELIABILITY - RANDOM FAILURES

The conventional method for the reliability analysis of electronic systems is a component level failure mode and effect analysis (FMEA) which involves assessing the potential effects on the system of each mode of failure of each component. Because of the large numbers of components involved, the internal complexity of the integrated circuit components themselves and the time-dependence of internal states in a programmable system, this approach is impractical for the fuelling machine control system.

The "fail-safe" and "fail-danger" modes of the system are well defined, the latter being associated with outputs that control plant drives. As these are normally open contacts, only spurious and maintained closure of one or more of these contacts could lead to an unsafe condition - i.e. a contact must be kept closed for a drive to be kept running as opposed to short term closure of contacts to initiate a drive and short term closure of other contacts to stop it. Hence, because these states are known, a "top down" approach is preferred in which failure rates of functional blocks within a sequencer are derived and the potential effects on the system of failed functional blocks are assessed as either "fail-safe" - i.e. not leading to spurious machine drive output closure - or "fail danger" i.e. leading to maintained closure of one or more machine drive outputs. Since precipitation of many of the hazards in the hazard schedule requires spurious closure of more than one output (e.g. a valve unlock actuator and a valve open drive must be energised in order to spuriously open the main valve on the fuelling machine), some pessimism has been introduced by basing analyses on logic faults which affect at least (and possibly only) one output.

The basic functional block definition is taken as the module (i.e. card) but is extended to encompass serial lines, output relays, power supply units, etc, as individual blocks. A module is said to have failed when one or more of the components, tracks, soldered joints or connections on that module have failed. Hence a module failure rate is calculated as the sum of the random failure rates of all its constituent parts. As a general rule it is assumed that failures which affect sequencer outputs are equally likely to result in spuriously closed or spuriously open outputs. The possible failure modes of the system are refined further by assessing, on a module by module basis, the maximum number of outputs that could be spuriously closed by any fault. When deriving the overall rate of faults affecting a particular output, faults which could lead to the simultaneous closure of a number of outputs are potentially more likely to affect the output under consideration and these factors must be included in the reliability analysis.

Standard component failure rates are scaled by multiplicative factors to account for the environmental conditions and lead to a rate for faults which could affect a particular drive output on channel 1 of 0.4 faults per "year of continuous control system operation". Assuming the system is used to handle 50 fuel assemblies per reactor year, each complete cycle taking up to 18 hours, this fault rate is equivalent to 4×10^{-2} per reactor year. For "fail-safe" modes the equivalent rate is calculated to be 10 per reactor year. These rates may be expected to decrease as the period of use builds up and failed components are replaced.

The overall channel probability of failure on demand is calculated (module by module) by combining the derived frequencies of "fail-danger" modes with the expected mean times to detection for faults on the appropriate module. This time to detection is determined from the frequency and assessed effectiveness of tests applied to the module. Three groups of tests enter into the calculation:

(i) Automatic self-tests and diagnostics. All have a maximum cycle time of 10s. On all cards containing a CPU the contents of ROM are checked, RAM operation and data lines to/from RAM are checked, serial lines are polled, a software maintained watchdog is refreshed, optical isolator inputs are forced and checked and, lastly, on-card output relays are pulsed and checked.

(ii) Start-up, pre-refuelling cycle tests. These are performed immediately prior to each fuel handling cycle (i.e. effectively with an 18 hour period) and comprise a check on the state of all plant inputs, execution of a software sequence which energises machine drive outputs and checks operation via the final contactor feed-back input, a check on all normally closed inputs by removing power from plant monitoring devices and, lastly, comprehensive memory checks.

(iii) Every 3 months, checking and maintenance is carried out on the fuelling machine and its control system. A range of tests is performed at this stage, designed to detect all equipment failures.

In general, self diagnostic tests are assumed to be 90% effective at revealing faults in the areas of circuitry which they cover. For each module, coverage by all relevant tests is taken into account to derive the percentages of faults revealed within 10s, within 18 hours and at 3 months. These percentages are combined to calculate the required mean times to fault detection.

The resulting net probability of failure on demand for the channel 1 logic is 9×10^{-5} and a similar result is obtained for channel 2 of the fuelling machine control system. Hence the combined failure probability of the two channels due to random hardware failures is calculated to be very low ($\sim 8 \times 10^{-9}$) and the frequency of coincident "fail-danger" states on channels 1 and 2 logic is calculated to be 3.6×10^{-5} per year of continuous operation. It is clear that these quantities must be dominated by common mode failure considerations and the general conclusion of this section is that in complex, programmable systems with well-defined failure modes, frequent, automatic self-checking and diagnostics can lead to the demonstration of a low overall probability of failure due to random hardware failures.

HARDWARE RELIABILITY - COMMON MODE FAILURES

The design of the safety systems described above has been strongly influenced by recognition of the importance of CMF. On the station in general it is recognised that where systems incorporate redundancy to achieve high reliability there may be a limit beyond which the probability of a common mode failure dominates. This results in the design requirement for diversity where very high reliabilities of protection are needed. Typically throughout the station, the reliability required of individual, non-diverse systems is a probability of failure on demand of no better than about 10^{-4} and it is judged that the achievement of this value will not be compromised by CMF. This judgement is made on the basis of the standards of engineering design and the formal approach to quality assurance adopted. In particular, this level of reliability is required from the combined channels 1 and 2 of the fuelling machine control system and, since the probability of failure on demand due to random hardware failures has been shown to be $\ll 10^{-4}$, CMF becomes a dominant aspect of the reliability analysis.

A systematic study of US nuclear and world aircraft common mode accidents has been made by the UKAEA Safety and Reliability Directorate (2,3). The CMF classification scheme and defensive strategies formulated therein have been followed in the development of fuel route control systems. In particular, aspects of the system structure, design methods, manufacturing methods, commissioning, maintenance,testing, operation and environmental conditions relevant to CMF have been investigated qualitatively as part of the design exercise. Since the nuclear protective subsystems covered in Ref 2 were typically hardwired guardline, input and output, the scope of the fuelling machine system investigation extended beyond that of Ref 2 by considering CMF aspects of programmable systems. CMF data relating to PES's is particularly scarce (for an example of that available see (4)) and hence the considerations were necessarily qualitative. In general it was concluded that the benefits afforded by the use of programmable systems (principally comprehensive and frequent automated self-testing, sophisticated indication and display interfaces and ease of maintenance due to automated fault diagnosis) were sufficient to offset the inherent complexity of these systems in terms of the risk of common mode failures causing simultaneous "fail-danger" conditions on two redundant channels.

In numerical terms a total CMF rate of 10^{-2}/year was assessed for the fuelling machine control system channels 1 and 2 together. For protection against any individual hazard (which in general represents only a fraction of the duty of the system as a whole) a CMF rate of 10^{-3}/year was assessed. The relationship to a CMF probability of failure on demand is made in terms of the expected mean times to detection of CMFs. Refs 2,3 did not specifically report the duration of CMFs but assumed an average duration of 10^{-1} years based upon typical proof test intervals of between 1 and 3 months for conventional systems. Although much more rapid detection by the self test and diagnostic facilities might be expected on the fuelling machine system, the same duration was assumed. Hence the assessed probability of failure on demand of channels 1 and 2 together was 10^{-4} to be applied to individual hazards.

SOFTWARE RELIABILITY

It is considered that the concept of error-free software cannot currently be realised in any but the simplest of programs and hence measures to reduce the possible effects of software errors are taken. The design intent adopted is that no single software error is to invalidate more than one line of protective logic. Diverse software production methods between channels are specified in order to meet the intent. If achieved, this reduces the possibility of software induced CMF to the level of random, coincident residual faults. In a numerical sense the design intent is that software faults shall not compromise the assessed CMF reliability measures discussed in the previous section. Since quantitative demonstration of achievement cannot be made qualitative aspects of the methods of software production, discussed briefly below, are assessed and lead to the judgement that dependent, software faults in both channels leading to a potentially unsafe condition can be discounted.

The software for each channel is coded in a proprietary high level language which was developed specifically for sequence control and interlocking applications. The code is a direct implementation of sequence flow charts which are themselves designed from schedules of interlocks and operations. As the interlocks are specified in order to meet the reliability requirements the software is a direct descendant from those requirements.

Production diversity is implemented down to the level of source documents for the preparation of flow charts - the documents for each channel are produced by different teams. Any common reference documents used to produce the flow chart source documents are subject to independent verification by teams within different organisations. When they have been produced, the flow charts are checked against source and reference documents by independent organisations and are subject to regular design reviews throughout subsequent stages of the software life cycle. There are several stages in the production of code from the verified flow charts (illustrated schematically in Figure 5), each with its own quality control documents defining manner of progress and fault recording, working procedures and methods for the implementation of modifications. Sequences are coded in a standard format with a layout designed for clarity and readability. After successful compilation each sequence is tested on a single rack test-bed system with real I/O references replaced by test inputs from local switches and LED outputs. All paths in the flow chart are covered and all tests independently witnessed. All aspects of the production cycle are carried out independently for channel 1 and channel 2 software before combined, interactive tests are performed.

Figure 5 Software Production Schematic

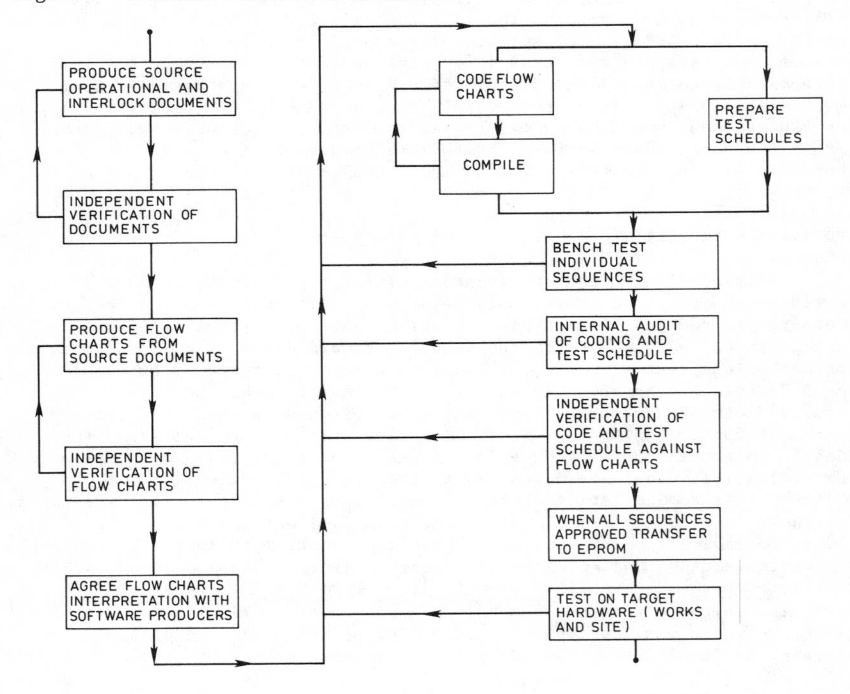

CONTROL SYSTEM RELIABILITY ASSESSMENT SUMMARY

The four independent channels of the fuelling machine control system comprise two based on programmable logic and two hard-wired channels providing selective diversity for interlocks where very high reliability is required. The discussion above has concentrated on assessment methods applied to the programmable channels. A "top-down" random failure modes analysis of the logic hardware results in a combined, 2-channel failure probability on demand of $\sim 10^{-8}$, a very low value due principally to the frequent, rapid automatic self-testing features incorporated. In the absence of any established method for a quantitative reliability assessment of the software element, the highest standards of practice were applied to specification, design, coding, documentation, testing and verification activities in addition to the implementation of diverse production techniques between channel 1 and channel 2 software in order to meet the design intent that no single software error is to invalidate more than one line of protective logic. Hence the channel 1 and 2 combined reliability is limited by common mode effects and a detailed, qualitative CMF assessment was performed (by two independent organisations) against guidelines provided by an independent study (2,3) of these phenomena. The numerical recommendations of the independent study were extended to cover the two programmable channels under consideration with the conclusion that the design target of a failure probability on demand of 10^{-2} for each channel - i.e. 10^{-4} for channels 1 and 2 combined - is not compromised by CMF effects.

CONCLUSIONS

A standard, modular PES has been incorporated into two out of four channels of a high integrity protection system. It has allowed the compexities of functional control of a significant number of plant drives and indications to be combined with adherence to stringent safety requirements on operations which carry a risk of leading to a serious hazard. A clear statement of the safety requirements forms the basis for the development of a hazard schedule, and a preliminary hazard analysis ensures that a consistent set of protection interlocks are provided covering a range of reliability requirements. The resulting interlock schedule determines the coarse structure of the protection system and a reliability assessment covering hardware random failures, common mode failures and software failures is performed to demonstrate in detail that the system meets its safety specification.

REFERENCES

1. Pressurised Water Reactor Design Safety Guidelines, CEGB Report DSG2, Issue A, 1982

2. Edwards, G.T., Watson, I.A., A Study of Common-mode Failures, UKAEA Safety and Reliability Directorate Report SRD R146, 1979

3. Bourne, A.J., Edwards, G.T., Hunns, D.M., Poulter, D.R., Watson, I.A., Defences Against Common-mode Failures in Redundancy Systems, UKAEA Safety and Reliability Directorate Report SRD R196, 1981

4. Wright, R.I., Some Data on Common Cause Failures in Redundancy Industrial Computer Systems, The Nuclear Engineer 26 No 3, 72

GUIDELINES FOR THE SYNTHESIS OF SOFTWARE FOR DISTRIBUTED PROCESSORS

CARPENTER, G.F., TYRRELL, A.M. and HOLDING, D.J.

Department of Electrical and Electronic Engineering
and Applied Physics
Aston University
Birmingham

ABSTRACT

A system of distributed processors offers an attractive method for the control of many real-world systems, with the prospect of increased efficiency, throughput and reliability. Modern software engineering analysis methods, design techniques and programming languages should be used in the construction of such systems to control and exploit the parallel nature of the system. Where a robust system is required, particular attention must be paid to the role of interprocess communications, because they provide not only a mechanism for synchronising and co-ordinating the distributed system, but also a mechanism for the propagation of errors. A proper fault tolerant framework must be implemented to restrict such error propagation and to provide proper conversation error-recovery mechanisms.

INTRODUCTION

Microprocessors now offer high computational power, high reliability and low power consumption at a low cost. They are finding widespread use in instrumentation and control systems where the microprocessor provides a centralised computing resource. Increasingly, microprocessors are being used in the construction of decentralised and distributed systems, in which a number of processors are physically distributed about the application plant and interact, or exchange information, with each other by passing messages over interprocessor communication channels. The individual processors in these systems not only provide local functions, such as data acquisition, control, and operator interfaces, but also form part of an overall system which must be co-ordinated to give a global response.

The primary concern in the construction of a computational system is to produce a design which satisfies the requirements specification of the system. The question of whether a computing resource should be implemented as a centralised or distributed system may be only of secondary importance. When a satisfactory design has been generated, and a computational architecture selected, a software specification for the chosen system has to be drawn up. Whether the computational parts of the system be centralised or decentralised, the function is determined by the software. Software must therefore be designed which meets the software specification and this design must be converted accurately into a program implemented on the target processing system. The resulting system will only operate correctly if the software is properly designed. For practical applications, the production of correct software is non-trivial.

The requirements specifications of monitoring and control systems often demand high levels of performance from a computational system. For example, the computational task may involve real-world data acquisition, combinational or sequential logic functions, complex arithmetic calculations, and the generation of control outputs to the application plant. The computational response may be required within very tight time constraints, perhaps as part of a real-time schedule. The schedule may have to be maintained in the presence of asynchronous external inputs, such as operator commands or alarms. In addition, the system may have to perform safety functions or functions with safety implications.

Requirements of this type make severe demands on the software, both at a systems level (involving the control and allocation of processor resources), and at the application level (responsible for the control of the plant). The design of such systems requires a proper understanding and application of the appropriate design techniques. These include, in the case of distributed systems, methods for the design of concurrent processing systems. The quality of the software, and of the resultant system, is critically dependent upon the adoption of proper methods and disciplines throughout the software life cycle (1).

This paper addresses some of the problems involved in the design of software for distributed processors, particularly where there are implications for safety. Modern software engineering techniques and languages are used to consider possible approaches to the design of such systems, and to discuss methods of providing fault tolerant structures for high reliability applications.

DESIGN CONSIDERATIONS FOR DISTRIBUTED PROCESSES.

The requirements specification for a computer system is chiefly concerned with identifying the functions which the system has to perform, the interfaces with the plant, and constraints within which it must operate. At this stage it is unlikely that a definite need to decentralise the computational system, or to distribute it, will have been identified. Indeed, only a detailed analysis of the requirements may lead to the decision that a distributed system is appropriate. The decision will normally be based on the following characteristics:

- Functional distribution

A real world system may be naturally distributed in a functional sense. Functionally distributed systems are often modelled and controlled as a set of distributed processes. The software for such a system invariably reflects the distributed nature of the application. This should provide a good correspondence between the real-world system function and the computational function.

- Geographical distribution

Real world systems are often spatially, or geographically, distributed. It is then appropriate to distribute the computational resource across the plant, and to design software which can be implemented over the set of physically distributed processors. Such software will necessarily consist of a set of communicating processes. Since many geographically distributed applications also have functionally distributed attributes, then both characteristics naturally lead to software designs which consist of a set of communicating distributed processes.

Once the decision to distribute the system is taken, then the software design and synthesis must adopt design rules and techniques which will lead to a high probability of generating correct, properly validated code within the specific demands of a distributed system (2).

- Partitioning and the reduction of complexity

The technique of partitioning is used to divide a system into a set of processes. The criteria used to partition a system can alter the extent to which interprocess communications are necessary to maintain the overall system function. System partitions are often chosen to emphasise the physical topology of the plant, the functional characteristics of a system, or the physical location of the processors. If they are chosen so that they emphasise the dominant characteristics of a system, they may give, to a first approximation, a fully decoupled system.

In many cases the partitions lead to an apparent reduction in the complexity of the system, or allow aggregation to reduce design complexity. However, the granularity introduced by partitioning should be carefully considered because it will affect the type of system implementation. For example, as the number of parallel processes into which a computational task is partitioned is increased, so the volume of interprocess communications for control and data interchange is also increased, thus leading to a closely coupled system implementation.

- Concurrency

Physical processes in continuous plant inherently involve the flow of energy or materials which often flow simultaneously through parallel forward paths, or forward and feedback paths. When such systems are modelled, the parallel processes are represented by parallel or concurrent data flows and are readily amenable to parallel processing for model simulation or control. This removes the constraint of modelling these systems in sequential terms which is required for solution by sequential computer programs executing on computers with a Von Neumann architecture. Concurrent programming languages and computing systems can therefore be regarded as the digital equivalent of the analogue computers, simulators and control systems which find widespread acceptance and continuous use in industry.

DISTRIBUTED PROCESSING

Each process (or processor) in a spatially or functionally distributed system may be equipped with local data acquisition or control interfaces. If each process is operated independently without communications with other processes then the system is said to be decoupled and each process can only operate asynchronously and autonomously and execute its local function only. Unfortunately, few practical applications have the characteristics necessary for decoupled control.

If a system can be controlled using a network of communicating processes, then the system is said to be coupled. The volume of interprocess communications determines the degree of coupling. In a loosely coupled system, relatively infrequent interprocess communications can be used to compute partitioned functions or to co-ordinate the distributed processes. A system is said to be closely or tightly coupled if there is a closer coupling between the component processes such that a high degree of interprocess communication is required to control and co-ordinate the system. Since the availability of communication links is often limited and the bandwidth of such links decreases with distance, closely coupled systems are often implemented as sets of processes communicating through shared memory on a centralised single or multiprocessor computing resource. Loosely coupled systems on the other hand can easily be implemented as geographically or spatially distributed systems.

A distributed system is said to be decentralised if the distributed processes have incomplete and non-identical information about the system state. Such a system requires the co-operative action of constituent processes in order to provide total system observability, controllability and overall function. The distribution of system function and the decentralisation of information can be used to enhance the robustness of the system.

For example, if the system is designed with the ability to recognise failure and can identify the processor or process concerned, then it may be possible to contain and isolate the fault. In distributed systems error migration through communications is a particular problem and it may be necessary to backtrack and trace or limit the effect of the erroneous communications. The reliability of the system may also be increased by the use of fault recovery techniques. If the fault leads to decreased functionality, then it may even be possible to regenerate a degraded function using other processes or processors provided the surviving communication systems will support the communications necessary for the recovery and operation of the reconfigured system (3).

COMMUNICATING SEQUENTIAL PROCESSES (CSP)

The software design of distributed systems necessarily involves the design of a set of communicating sequential processes, involving aspects of concurrency. The methods for the design of centralised multiprocessing systems have been developed over the last eighteen years (4-6).

However, the techniques for the identification of critical sections of code, and the provision of mechanisms for enforcing mutual exclusion and synchronism essentially provide bottom-up design primitives. They are used extensively in the kernal of design methodologies such as MASCOT (7) and are hidden from the applications designer. Such monitor based techniques are unsuited to distributed systems design since a centralised facility is unavailable.

The development of concurrent programming languages, such as CSP and its derivatives, such as occam (8-9), in which message-passing synchronising inter-process communications are a primitive of the language, allows the high level design of distributed systems. The use of such constructs simplifies systems analysis and facilitates the design of distributed systems. The formal background of CSP also provides an mathematical basis for the analysis of the system behaviour and the design of fault tolerant methods.

A CLASSIFICATION OF INTER-PROCESS COMMUNICATIONS.

Interprocess communications may be classified (10) into one of three groups:

i) synchronous communications, where neither the sending nor receiving process is allowed to proceed beyond the communication point until its complementary process has also reached that point. This is found most notably within CSP and occam.

ii) asynchronous communications, where the process sending the message does not wait for acknowledgement, but the receiving process is not permitted to proceed beyond the communication point until the message has arrived (11).

iii) remote procedure call, where the process sending the message requires the receiving process to perform some specific function and respond before they both can proceed further. In essence it is an asynchronous communication followed by a synchronous communication. This form is found in ADA and elsewhere (12-13).

It is common to associate inter-process messages with the function they perform (10). Alarm messages have high priority on the interprocess medium; they are issued by one process and require immediate response by the receiving process. Command messages require a change of state or action to occur in due course; acknowledgement is a necessary requirement before the issuing process continues. Status messages are sent to notify other processes of information about the source task. No acknowledgement is required.

These message groups can be constructed using any of the above communication primitives. However, detailed study is required to ensure that the logical structure of the inter-process action (the transaction level) is not disrupted should the lower level communication primitives fail in a particular application (14). Single failure detection systems, such as time-outs, can only be applied at process level on a per-process basis and this does not necessarily provide protection at the transaction level.

For example, failure in an asynchronous communication system can leave a process suspended indefinitely awaiting a communication, or may leave one process aware of a failure but unable to co-ordinate recovery action through the absence of a logical pairing between participating processes. The remote procedure call requires the object task to acknowledge when its action is complete (12); if the reply is lost then the source process is suspended indefinitely. Synchronous primitives, such as those in occam, preclude the protection of individual transactions since the language is specifically intended for deterministic system design only and outputs can not be used as guards on synchronous primitives. Hence, timeouts cannot be used in parallel with other processes to form a race condition, and so they cannot be placed on the send primitives of inter-process transactions. However, experimental languages, such as Pascal m (15), have attempted to overcome these deficiencies, but no general consensus nor formal method is as yet suitable. Although transaction level protection cannot be supplied directly, software fault tolerance methods can be applied to such systems using state based recovery techniques which may enclose complete inter-process communication transactions within the distributed recovery block or conversation (16).

SOFTWARE DESIGN TOOLS FOR DISTRIBUTED SYSTEMS

The design and synthesis of software for distributed systems requires the use of a design methodology and programming language which builds on the inherent parallel nature of such systems. Formal methods applied to the design of software for distributed processors lead to the identification of processes, capable of asynchronous execution, interacting with other processes by communications. The provision of constructs for sequence, variable assignment, selection and iteration, augmented by constructs which enable parallel execution, the use of synchronous communications for input/output, the provision of guarded processes and the formal inclusion of time are sufficient for the design of software for distributed systems (17-18).

The programming language occam, which is derived from the theory of communicating sequential processes provides a good notation in which to pursue the design of distributed systems. Occam produces concise, elegant and easily understood software. The mathematical axiomatic base of CSP permits algebraic analysis of software. In particular, correctness preserving transforms can be applied to it. It therefore offers the prospect that in the future such software may be formally verified.

The language is intended for use with both sequential and inherently parallel systems. The starting point in design using occam is the identification of the natural parallelism and the partitioning of the software into naturally occurring processes. Interaction between processes is solely by means of interprocess communication. Again this is a good match with a distributed system. The mapping of processes onto target processors occurs at a late stage in the design. This allows the designer to concentrate on the application function rather than implementation details.

HIGH RELIABILITY DISTRIBUTED SYSTEMS

For high reliability applications it is essential that failure modes be identified and measures taken to ensure that the system recognises when a fault occurs, constrains the scope for error propagation, and recovers to generate a safe response (19). A taxonomy of faults, ranging from sensor failure to software faults, can be drawn up with methods for their detection, and appropriate remedial action. The framework for recovery is relatively straightforward for centralised sequential systems, involving the use of process roll-back within recovery blocks and offering alternative processes (20), possibly within the confines of time-out watchdog timers (21). A fundamental assumption is that the framework itself is immune from faults.

However, the situation becomes more complex for decentralised systems since there is scope for error propagation by inter-process communication, which once initiated cannot be retracted. Methods are required which restrict the scope of error propagation between communicating processes, and which co-ordinate recovery amongst all processes participating in erroneous communications. The conversation scheme (20) is appropriate, and requires all participating processes to perform an acceptance test at pre-determined points in their processing. If any process is found to be in error at that test then all participating processes must perform co-ordinated recovery. No process is allowed to proceed beyond the acceptance test until all the other processes also pass the test.

The chief design problem is the proper placement of conversations. It is evident that the conversation scheme requires synchronism at, or following, the acceptance test to exchange the results of the test and, if necessary, coordinate process action. Current approaches to this problem have used the centralised concept of a monitor to implement acceptance tests. Unfortunately, this centralising feature of the monitor makes it unsuitable for distributed systems. An alternative approach developed by the authors (16) makes use of the synchronising properties of CSP/occam communications to design and implement distributed acceptance tests.

The crucial problem of conversation placement has received somewhat less attention. In effect the designer must identify the extent of process corruption and error migration through inter-process communication for all faults in the system. The objective of this fault effect analysis is the identification of a boundary or set of properly nested boundaries, which define known entry (recovery line) and exit (acceptance) states for the system. This allows the entry and exit state for each component process to be determined. Software must be designed to save recovery line entry states, and to implement and synchronise the acceptance tests on all processes in the conversation. Attempts to identify recovery lines and acceptance points dynamically are prone to progressive collapse (20).

An alternative approach (16) is to utilise the deterministic state properties of CSP/occam in the static design of conversations within the known state reachability space of the distributed system. This approach offers considerable advantages and allows the use of design aids which automatically generate sets of proper conversation boundaries within the system. It is then for the designer to choose the features of a design which he wishes to protect and the degree of software fault tolerance appropriate to a particular class of application.

The conversation scheme offers the most appropriate structure for recovering from unanticipated faults. The nature of the conversation scheme is that the acceptance test results (go/no-go) must be compared amongst the participating processes. The detection of an error during an acceptance test does not necessarily identify uniquely the fault; indeed the fault might lie in the interprocessor communication medium. Circumstances may arise where it is impossible to promulgate the result of the acceptance test to promote error recovery, perhaps due to a failure of the interprocess medium. In this cases the framework for recovery fails because the fault affects the recovery structure itself. Similar problems are inherent in any recovery structure and have been recognised for recovery block structures applied to sequential software (22). However, thay have not detracted significantly from the effectiveness of the technique.

A CONTROL EXAMPLE

The program fragment presented in Figures 1-4 is taken from a example program used to explore the problems involved in the design of software for distributed processes. It concerns the motion of a robot in each of three axes. It illustrates a number of points:

i) inherently local functions are modelled as processes, each capable of execution in parallel. Thus motion in each axis is programmed, and occurs, independently of motion in the other axes. Each process could be targetted onto separate processors at a late stage in the design.

ii) interprocess communications is the only form of interaction between the processes. This would take place over an interprocess communication medium. Thus each process is commanded to perform its activity, and signals when it has completed its activity.

iii) The software also contains a command process which initiates parallel commands and receives, as they occur and in whatever order they occur, the responses corresponding to the execution of those commands.

iv) The software contains a proper conversation boundary for the protection of the critical interprocess communication which governs the co-ordinated axial movement of the robot.

An equivalent fault-tolerant program written in a conventional language would be much more difficult to design, program and verify.

Figure 1

```
{{{ PROGRAM robot
... system parameters
CHAN request, return, motion[3], finished[3], stop[4]:
... process operator
... process motor
... process control
-- initiate processes
PAR
  PAR i = [0 FOR 3]
    motor(motion[i], finished[i], stop[i])
  control(request, return, stop[3])
  operator(request, return)
}}}
```

Figure 2

```
{{{ process motor
PROC motor (CHAN motion, finished, stopi)=
  VAR step, direction, going:
  SEQ
    going := TRUE
    WHILE going
      ALT
        stopi ? ANY
          going := FALSE
        motion ? step
          SEQ
            motion ? direction
--  move motor
            finished ! ANY:
}}}
```

Figure 3

```
{{{ process operator
PROC operator (CHAN send, receive)=
  VAR x, y, z, run:
  SEQ
    run := TRUE
    WHILE run
      SEQ
        Screen ! 'i'
        Screen ! EndBuffer
        Keyboard ? x
        Screen ! x
        Screen ! EndBuffer
        Keyboard ? y
        Screen ! y
        Screen ! EndBuffer
        Keyboard ? z
        Screen ! z
        Screen ! EndBuffer
        x := x - '0'
        y := y - '0'
        z := z - '0'
        IF
          (x=0) AND (y=0) AND (z=0)
            SEQ
              Screen ! 'f'
              Screen ! EndBuffer
              PAR i = [0 FOR 4]
                stop[i] ! ANY
              run := FALSE
            TRUE
              SEQ
                send ! x
                send ! y
                send ! z
                receive ? ANY
                Screen ! 'm'
                Screen ! EndBuffer:
}}}
```

Figure 4

```
{{{ process control
PROC control (CHAN receive, send, stopi)=
  VAR xold, yold, zold, xnew, ynew, znew, count, step[3], direction[3],
      going:
  SEQ
    xold := 0
    yold := 0
    zold := 0
    going := TRUE
    WHILE going
      ALT
        stopi ? ANY
          going := FALSE
        receive ? xnew
          SEQ
            receive ? ynew
            receive ? znew
-- calculate distance and direction of each
-- motor. These can be calculated in parallel.
            PAR i = [0 FOR 3]
              SEQ
                motion[i] ! step[i]
                motion[i] ! direction[i]
            xold := xnew
            yold := ynew
            zold := znew
            count := 0
            WHILE count <> 3
              ALT i = [0 FOR 3]
                finished[i] ? ANY
                  count := count + 1
            send ! ANY:
}}}
```

CONCLUSION

The design of distributed computer systems requires specific methodologies and techniques if high reliability is to be achieved. Systematic analysis of the specification is required to identify and to exploit the parallelism inherent in the application. This must be complemented by design methods and programming languages suited to a highly parallel computing environment. Careful analysis of the communications is required to co-ordinate the processing. and to ensure that proper conversations are produced for recovery activity.

REFERENCES

1. Daniels, B.K., Reliability Engineering, 4, 1983, 199.

2. Boehm, B.W., Software Engineering Economics, 1981, Prentice-Hall.

3. Momen, S.E.M., Holding, D.J., Proc Int Conf on Control and its applications, I.E.E., 1981, 291.

4. Brinch Hansen, P., Operating System Principles, 1973, Prentice-Hall.

5. Dijsktra, E.W., Co-operating sequential processes, 1968, in Programming Languages, ed. Genuys, F., Academic Press.

6. Hoare, C.A.R., Comm ACM, 17, 1974, 549.

7. Simpson, H.R., MASCOT 3, I.E.E. Coll on MASCOT, 1984.

8. Hoare, C.A.R., Comm ACM, 21, 1978, 666.

9. May, D., Sigplan Notices 18, 1983, 67.

10. Kramer, J., Magee, J., Sloman, M., Proc 2nd Int Conf on Distributed Computer Systems, 1981, 404.

11. Liskov, B., Proc 7th ACM SIGOPS Symp on Operating System Principles, 1979, 33.

12. US Department of Defense. ADA Reference Manual. 1980.

13. Kramer, J., Magee, J., Sloman, M., Lister, A., IEE Proc Pt E, 130, 1983, 1.

14. Holding, D.J., Carpenter,G.F., Tyrrell,A.M., Proc 6th IEEE/Eurel Conf on Computers in communications and control, 1984, 235.

15. Bornat, R., A protocol for generalised occam, Research report 348. Queen Mary College, 1984.

16. Tyrrell, A.M., Holding, D.J., submitted to I.E.E.E. Trans on Software Engineering, 1986.

17. Linger, R.C., Mills, H.D., Witt, R.C., Structured programming, theory and practice, Addison Wesley, 1979.

18. Hoare, C.A.R., Communicating Sequential Processes. Prentice-Hall; 1985.

19. Anderson, T., Lee, P.A., Fault Tolerance: Principles and Practice, 1981, Prentice Hall.

20. Randell, B.R., I.E.E.E. Trans on Software Engineering, SE-1, 1975, 220.

21. Kim, K.J., I.E.E.E. Trans on Software Engineering, SE-8, 1982, 189.

22. Jackson, P.R., White, B.A., The application of fault tolerant techniques to a real-time system., Safety of Computer Control Systems, Pergamon, 1983.

EXPERIENCES WITH THE DIVERSE REDUNDANCY IN PROGRAMMABLE ELECTRONIC SYSTEMS (PES)

by W. GRIGULEWITSCH, K. MEFFERT and G. REUSS

INTRODUCTION

In "Berufsgenossenschaftliches Institut für Arbeitssicherheit - BIA -" diversely redundant machine control systems were tested under safety technique aspects. The generally applicable experiences and information obtained by this investigation are described.

Prior to the description of technical details, several important marginal suppositions and conditions of the tested system and the failures to be considered have to be pointed out. The systems show the following common features:

- machines with dangerous movements are concerned requiring regular intervention in the dangerous zone;

- the machines have a safe state which is identical to a machine stop;

- the dangerous movements are linear, the movement can be induced by linear drive or by excenter;

- in this connection, controls - not automatic controls - are concerned, i.e. there is no feedback of controller signals in the sense of an automatic control technique;

- signal processing is binary.

For the investigation of a safety concept it is important to determine which kind of dangerous situations have to be handled in case of errors. The described tested systems must not include the risk of unexpected dangerous movements in case of failures. Furthermore a failure in the control system must be recognized within an operation cycle and should have the effect of preventing further operations. If failures cannot be detected in this way unnoticed failures may accumulate (failure accumulation). For testing purposes, up to three independent failures are assumed.

Accidental failures, defects of external lines and systematic failures and errors are regarded here. These failures are not described in this paper, they are included in /1/, however. Several failures, especially systematic ones, can be prevented by appropriate measures, e.g. by appropriate testing procedures. Different failures, mainly accidental failures, cannot be prevented. They must be kept under control during operation, therefore.

The question which kind of failures or faults in programmable electronic systems (in this case a microprocessor) can be assumed is of special interest. Supposing that software cannot be checked for failures up to a level of 100 per cent and functional effects of hardware failures are unpredictable the worst case is assumed: i.e. a total failure of PES. The worst case shows correct control routines and tests but wrong functional routines. This worst case assumption automatically leads to concepts with diverse redundancy.

THE DIVERSE REDUNDANCY PRINCIPLE

Fig. 1 as block diagram shows the principle circuit scheme of a diverse, redundant control structure. At the same time, the scheme shows the minimum system configuration for the selection of a single dangerous movement. The peripheral signals are supplied to two channels having a structure with different principles (diverse). Each of the channels K 1 and K 2 includes the complete functional logic of the machine and would be able to control the machine function on its own. The output signals of the channels A 1 and A 2 controlling dangerous movements are led to a comparator (V) and to an AND-circuit (&). The comparator is intended to check output signals for equivalence. In case of an antivalence, i.e. A 1 $\neq$ A 2, there is either an error in the periphery, or one of the channels has determinated a wrong result for the function due to an internal failure. Since it is unknown which of the channels produces the correct result there is no other possibility but to assume the safe state of the system via signal technique. This is the state "0" (stop of movement). The stop is realized by the comparator which sets "0" to the third input of the AND-circuit. Between comparator and AND-circuit a time function element is wired which is intended to suppress only shorttime antivalences of the channels K 1 and K 2. The causes of shorttime antivalences and the dimensioning of this time function element are discussed later.

PERIPHERAL COMMAND SWITCHES

Each control rated for a concrete application requires a certain number of command switches in order to secure, in addition to the normal function, the protection of the operator. Since the majority of control failures occur in components of the periphery or in their supplies it is important to control these failures by appropriate measures. If the signals for both diverse control channels are originated by a common command switch a failure of this system may result in a simultaneous influence of both channels unless appropriate isolating measures have been taken.

In principle, accidental and systematic failures can be controlled if signal results are diversely redundant and the check-up of single signals is realized by the control for equivalence or antivalence (valence check-up). Redundancy in combination with a valence comparison is required if a failsafe technique is not available, early detection of a failure is necessary, however. Diverse signal results mean that command switches should differentiate with regard to technology (construction element diversity) but also regarding the physical principle of actuation (acutation diversity) (fig. 2).

The follow-up control of peripheral command switches can only execute an equivalence or antivalence check if each channel receives the same signal contents. If processing channels merely receive parts of the signals only the follow-up comparator of the channels is capable of executing this valence check (fig. 3a). The circuit of fig. 3b is preferable to that of 3a, since the functional logic in the channels permits an early recognition of failures.

If peripheral signal lines are simultaneously connected to both channels it has to be considered that in principle a common system bearing the risk of possible systematic failures is concerned. By appropriate isolating measures can be ascertained that this channel connection is free of feedbacks, thus preventing that errors in the input part of a channel affect the other channel.

CHANNEL STRUCTURE, SIGNAL PROCESSING

The essential feature of a diversely redundant concept is the application of the diverse principle on functional channels. As already mentioned, diverse channels are necessary due to the worst-case assumption, i.e. that software cannot be checked up to a level of 100 per cent and the effects of hardware failures are unpredictable. Regarding the expenses of development, these failure assumptions not only show disadvantageous consequences but also have considerable advantages: a sufficient degreee of diversity for example may result in reducing the expenses for the program testing to an absolute minimum. People who know about the required times for software checks are able to avaluate the extent of these reductions. Under certain circumstances, several automatic test procedures within the channels can be omitted. In the following paragraphs, different aspects are discussed which are important for the design of diverse channels.

DEGREE OF DIVERSITY

The more different channels are designed the less probable it becomes that summing-up accidental faults or systematic failures are, due to a single cause, expressed in the same way in both channels and thus cannot be detected by the follow-up comparator. A high degree of different design principles is obtained if, for instance, one channel is operated by a microprocessor, the other one by a permanently wired logic (see fig. 4). If both channels are to be equipped with computers these have to be different (see fig. 5). The diversity can be raised further by appropriate measures:

- application of different algorithms for result calculation;

- different storage area engagement;

- technological differences of processors and of peripheral elements;

- different supply voltages;

- if possible, a different processor clock.

PES INTELLIGENCE UTILIZATION

The enormous flexibility of process computers can also be utilized to apply the available "intelligence" for valence checks, reasonableness inquiries and automatic tests. Thus, processor failures or peripheral failures can be early detected, and an accumulation of failures is prevented. Furthermore, operational errors can be registered without difficulty by corresponding inquiries.

VALENCE CHECK

If all peripheral signals are supplied to both channels (fig. 3b), as a rule, each channel is capable of determining whether the corresponding signals are equivalent or antivalent. In this way, failures in the periphery can be detected rapidly and used for a turn-off. However, only accidental, but no systematic failures can be recognized in this way.

PLAUSIBILITY INQUIRIES

For plausibility inquiries, the input signals are at certain times compared with prescribed signal patterns. For instance, if variations occur in case of a correct computer channel the failure is peripheral, and the system can be stopped. In general, by plausibility inquiries only peripheral signals can be checked, since this test is based on an operational computer channel. A further possibility to check that data from the periphery are correct is to restrict the validity duration of incoming data and to have the validity checked by a seperate program loop /2/.

SELFTEST

To secure that the processing channel is still working the available system possibilities can be used to execute additional selftests. The intention is to detect hardware changes in the channel. Since complex systems do not permit a complete test, ROM and RAM tests are partly executed. Peripheral signals relevant for safety purposes, e.g. light barriers, can be checked in regular intervals for correct function by the computer. Under certain circumstances, dependent on system size and extent of signal processing the tests are executed during longer operational interruptions of the complete system. Individual possibilities to test RAM and ROM are described in /3/. Systematic failures, e.g. regarding program preparation, cannot be detected by selftests.

TESTING OF SYSTEM SPECIFICATION

By the above mentioned measures in connection with the redundant concept it is very probable to detect accidental faults in the hardware but also failures in the program in time. Failures occurring before programming, mainly failures in the specification, would be unnoticed, however. Since the system specification is in general the basis for the programming of both channels its testing should be executed with extreme care. According to our experience, specifications with a future program scope of < 1 KByte can be satisfactorily tested by a second independent team.

COMPARATOR

The machine controls tested in BIA up to now were equipped with hardware comparators in positive contact action technique relays or with discrete electronic components.

Comparators guided via contacts are chosen when the number of signal comparisons to be executed was low (one to two) and when, after a positive comparison result, energy for the following circuit had to be available. Fig. 6 shows the principle of a comparator. This figure shows how the interlocking of relays can be obtained by using break contacts in front of the respective other relay coil /4/. The time shift of two signals (A 1 and A 2) to each other is tolerated within strict limits, greater variations result in a permanent blockage of further processing. The failsafe reaction of the circuit is obtained by using appropriate relays with positive contact action so that any failure has an immediate inhibitory effect.

In dependance on the number of safety functions of a machine it is required to check a defined minimum number of signals after their logical integration in processing channels for their matching. For a short failure detecting time it is important to supply the comparator circuits not only with final but also with intermediate results. For the machine types in question with one to two dangerous movements and two to three safety devices (e.g. two-hand switching, safety light barrier) four to six of these comparator circuits have proven to be appropriate to obtain a sufficiently short error detecting time.

In general, the comparator can only detect failures in the processing channels if the signals to be compared are varying. As a rule, this dynamic of processing channels with a discrete or low integration technique is combined with the machine operating cycle. When using microprocessors in processing channels an additional high dynamic expansion of signals to be compared is possible. This possibility can substantially reduce the failure detection time.

Simple failsafe comparator circuits consisting of a few discrete electronic components only, are listed in the references /5, 6/. Fig. 7 shows a failsafe comparator for antivalence according to the principle of the circuit system described in /6/. Only if the signals A 1 and A 2 show antivalent potentials to each other the bridge-connected rectifier is able to amplify the test clock at its basis.

TIME FUNCTION ELEMENT

Due to mechanic tolerances, the signals from the periphery may show shorttime antivalences thus inducing function channels to signalize different process stages at their outputs. But also the diversity in the channels has the effect that the output results are supplied at different times. Due to the short reaction times of the comparison, a system turn-off would be the result as shows the feedback to the inputs of processing channels in fig. 1. This factor which reduces the availability of the controlling system requires technical measures of the circuit to tolerate shorttime inconsistencies between both channels. The maximum time for the suppression of an incompatability is mainly dependent on machine operation and handling. The determination of this time finally requires the consideration of reaction times of follow-up position elements and also, if necessary, the time required for the braking of moving masses. Technically, this problem can be solved by the use of simple time cycle circuits with monostable and retriggering characteristics which secure the turn-off command in case of failure. In addition, time inquiries in processing channels via the software are possible.

SELECTION BY AND ELEMENT

The results produced by the functional logic have to be forwarded to element positions. In order to forward the results of the channels and the comparator to element positions a failsafe AND element is required. AND circuits of this kind are known to have relays in a positive contact action technique but are also realized by discrete electronic components. Fig. 8 shows the circuit principle of a failsafe AND element. The principle of dynamic signal processing is applied here. The comparator disposes of the power for a voltage frequency conversion. The output signals A 1 and A 2 supply the power necessary for modulation. The transmitter secures that only dynamic signals (alternating voltage) can be forwarded. In any imaginable element failure, the shown circuit proceeds to a safe state.

POWER SUPPLY

Most crucial in a redundantly structured control system are the common lines and the devices commonly used. It is recommended, therefore, to choose a redundancy in the power supply, too. Only in this way can be ascertained that accidental failures in the power supply element or electromagnetic compatability interferences do not result in similar failures in both channels. Based on the circuit scheme in fia. 1, at least two separate voltage supplies have to be provided (fig. 9). Each channel is fed from different voltage sources. One of these sources feeds at the same time the comparator and the failsafe AND element. In case of a high efficiency requirement of position elements in the output part, a third efficient power voltage is frequently necessary.

RELIABILITY, EXPENSES

Practical experiences with diversely redundant electronic control systems could be obtained with paper cutting machines. One of these systems has been available for several years with a piece number of more than 10.000. It was observed that these machines are very insusceptible to malfunctions and furthermore they are serviceable. Compared to the former conventionally controlled machines, the number of malfunctions is substantially lower. Figures about the operational availability are not at our disposal, however.

The hardware expenses for both channels, the comparator and the interface amount to three to four Europe cards. The power supply is included in this figure. Some figures about the development of safe software systems can be found in the references /7/. According to these date, the expenses per command are between $ 30 and $ 700 corresponding to safety relevant requirements.

These data could not be confirmed regarding the described systems. According to the information of machine producers, the development expenses for software were not much higher than the expenses for the development of safety relevant software. This is reasonable since the testing of the real programming only plays a secondary role in diversely redundant concepts.

Based on previous experiences, the testing times for a complete system are about half a man year. Hardware testing requires the majority of it (70 per cent), especially periphery and comparator testing, concept assessment and software testing amount to a proportion of about 15 per cent each.

REFERENCES

/1/ Griguletwitsch, W., Meffert, K., Reuß, G.: Aufbau von Maschinensteuerungen mit diversitärer Redundanz. BIA-Report (in Vorbereitung)

/2/ Janning, W.: Zweikanalige Pressensicherheitssteuerung mit einem speicherprogrammierbaren Steuerungssystem. VDI-Z 122 (1980), Nr. 1/2, S. 17 ff.

/3/ Hölscher, H., Rader, J.: Mikrocomputer in der Sicherheitstechnik. Verlag TÜV-Rheinland, 1984.

/4/ Wienandts, W.: Sicherheitstechnische Anforderungen an elektronische Steuerungen. Die BG, März 1980, S. 220 - 224.

/5/ Bernhard, U., Cordes, D.: Steuerungen mit erhöhtem Sicherheitsgrad. Elektroniker, Elektronik-Fachteil (1976) Nr. 1, S. EL1 - EL4.

/6/ Lohmann, H.-J.: URTL-Schaltkreissystem U1 mit hoher Sicherheit und automatischer Fehlerdiagnose. Siemens-Z. 48 (1974), H. 7, S. 490 - 494.

/7/ N.N.: Kursunterlagen: Application of Computer Systems for Industrial Safety and Control. Organized by Oyez Scientific and Technical Services Limited.

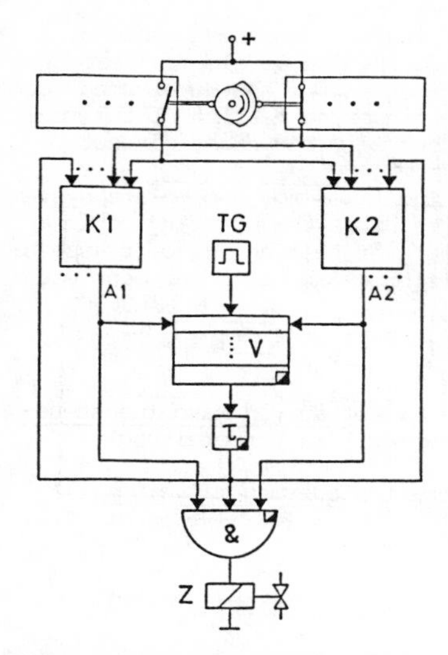

Figure 1: Circuit principle of diverse redundancy

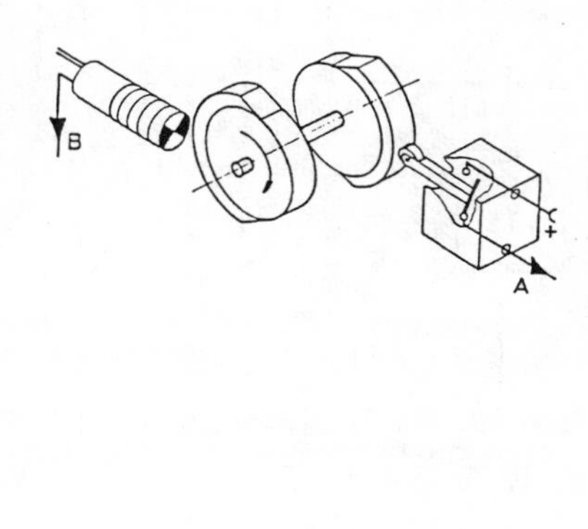

Figure 2: Diversity of components and actuation on approaching position switches

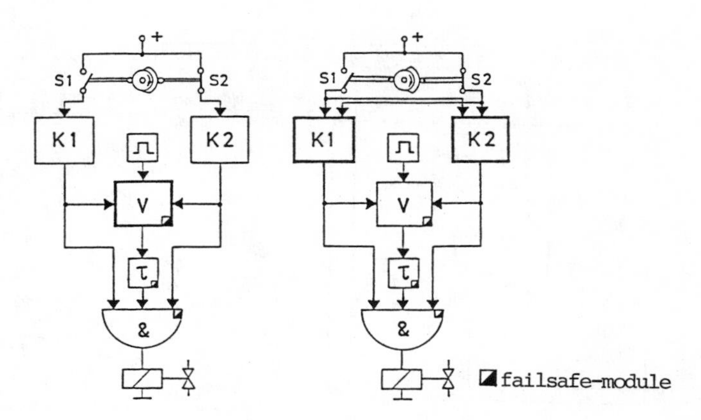

Figure 3a:

Figure 3b:

Testing possibilities for equivalence of a redundantly produced signal:

a) Valence comparison is only possible in comparator; isolation of channels has already been performed.

b) Valence comparison is only possible in channels and additionally in comparator; isolation measures have to be provided for.

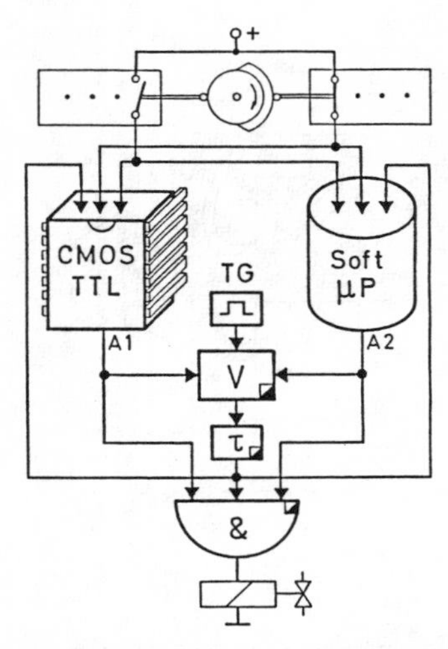

Figure 4: Realization of diverse redundancy with µP and CMOS/TTL-logic

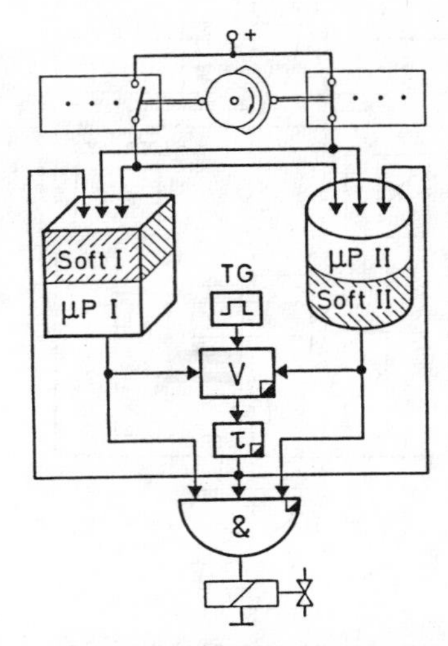

Figure 5: Realization of diverse redundancy with two µP

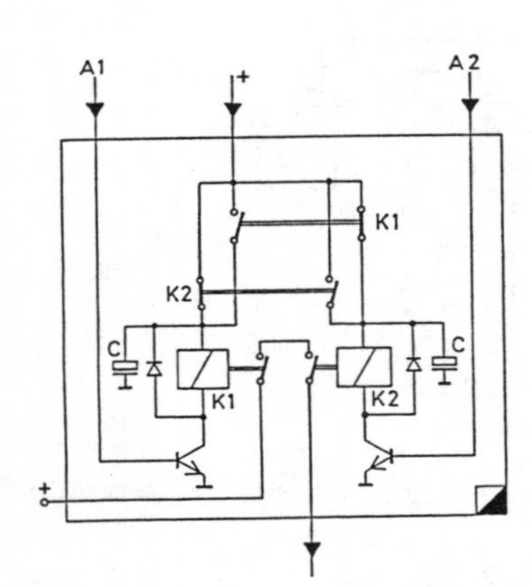

Figure 6: Circuit principle of a failsafe comparator with relays in positive contact action technique /4/

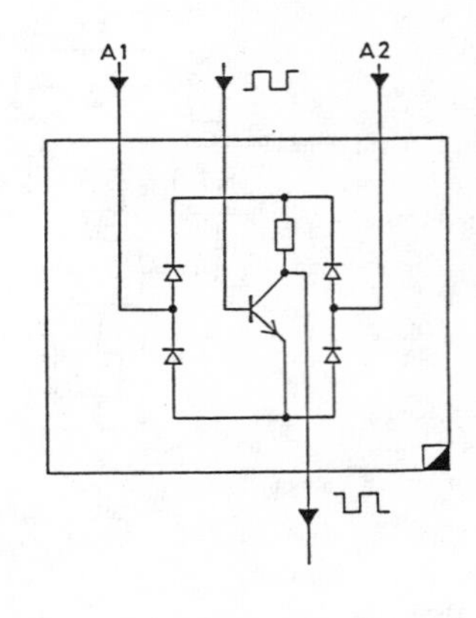

Figure 7: Circuit principle of a failsafe antivalence comparator realized by single electronic components

◪ failsafe-module

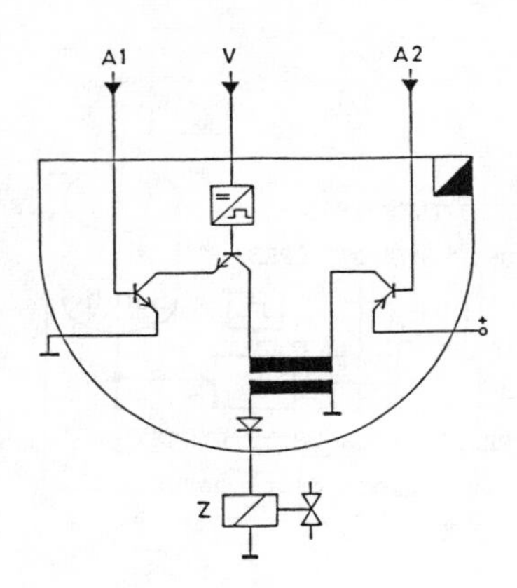

Figure 8: Circuit principle of a failsafe AND connection of diversely obtained output signals according to the principle of dynamic signal transmission via transformer

◪ failsafe-module

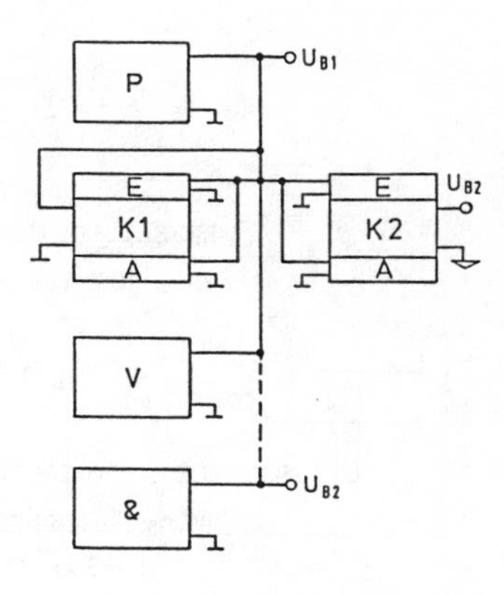

Figure 9: Control system supply by at least two separate voltages

EFFECTS OF ELECTROMAGNETIC INTERFERENCES ON PROGRAMMABLE ELECTRONIC SYSTEMS (PES)

B. HIGEL, D. DEI-SVALDI and B. CLAUZADE

INSTITUT NATIONAL DE RECHERCHE ET DE SECURITE (INRS)
Service "Electronique - Sécurité des Systèmes"
54501 VANDOEUVRE CEDEX - FRANCE

ABSTRACT

The susceptibility of programmable electronic systems (PES) to electromagnetic interferences (EMI) is known as an actual problem from the point of view of PES safety. At first, some specific aspects are focused on, f.i. those relevant to the programmable electronics of PESs.

Then, statistical results from a campaign of EMI tests conducted at INRS from 1983 to 1985 are presented. About 80 industrial programmable controllers (IPC) of different design, representative of equipment available on the world market, have been tested.

Some conclusions are drawn, and one point is emphasized : to conduct further studies it seems to be essential to explore the conditions of the electromagnetic environment actually encountered in industrial surroundings.

INTRODUCTION

The assessment of PES safety is one of the critical points that research concerning the safety of industrial automatized systems has to investigate (1). For the purpose, a specific methodology must be designed : the studies carried out on the topic agree on defining some key-steps the assessment process has to go through (see e.g. (2) and (3)). Not surprisingly, one of these steps is devoted to studying the effects physical environment can have on PESs.

The present paper deals with one feature of the physical environment which can be crucial for PES safety, i.e. the presence of conducted and radiated electromagnetic interferences (EMI). This topic is presently of growing interest, not only in the field of robotics safety but also in most of the applied fields of electronics.

The paper starts with some examples and some data which show electromagnetic compatibility (EMC) problems being relevant to PES safety. Then general aspects of PES safety with respect to EMC are focused on in a qualitative manner.

The most important Section is devoted to a presentation of quantitative results. They come from a measurement campaign worked out at INRS on the electromagnetic susceptibility of industrial programmable controllers (IPC). Two sets of data have been collected at a two years interval (1983 and 1985).

The Conclusion Section is a prospective discussion made on the further extension of these studies at INRS, particularly from the point of view of metrology of the electromagnetic disturbances actually found in industry.

PES SAFETY AND ELECTROMAGNETIC SUSCEPTIBILITY

EXAMPLES

The susceptibility of programmable electronics to conducted and radiated EMI is known as an actual problem as well as the risks which can proceed from. For example, a gas detector delivers an output signal for monitoring the atmospheric composition in an industrial plant. The processing electronic system triggers an alarm when a gas concentration level becomes larger than a given upper limit. Electromagnetic disturbances acting on the detector can develop an output signal misinterpreted by the alarm system. The power actuators of a robotic system are driven in a sudden random manner because of conducted or radiated interferences occuring in the processing unit or in the output interfaces. Locomotors are remote controlled through guided or free-space propagating signals : erratic dangerous movements can result from unexpected electromagnetic signals interfering with the communication system. A hand prothesis driven by electronic sensors "connected" to the forearm is closed in disabled position. It suddenly opens when a radiofrequency wave is transmitted in the nearness, e.g. by a walkie-talkie.

Very detailed and objective data are difficult to collect about that kind of problems and their consequences, because of several reasons : in addition to the confidentiality about accidents, their correlation with electromagnetic phenomena is often hard to establish. Nevertheless, some quantitative data are available.

For instance, a survey through mailed questionnaires has been carried out by INRS in 1984 on the anomalies stated during industrial operations of 362 robots installed in France (4). According to the received answers, the amount of robots working near devices able to generate radiated electromagnetic interferences appears to be 9 %. Frequent interferences in the power supplies alone are observed for 23 % of the robotic systems. From the general point of view of environmental conditions (i.e. EMI included), their influence on the robot behaviour is recorded in 31 % of cases, with occurrence of malfunctions (71 % of these cases), of data losses (42 %), and of other failures like breakdown of components (35 %).

ELECTROMAGNETIC SUSCEPTIBILITY OF PES's

The study of the interaction between any system and its physical environment usually requires to consider a number of aspects : environmental description, characteristics of the environment-system interface, modes of action, effects and consequences, and so on... In the present case of EM - PES interaction, exhaustive investigations regarding all these aspects would extend beyond the limits of that paper. Therefore only some specific points will be considered. The next Section, which deals with experimental results, will provide the opportunity to take up some other ones.

A PES can be considered as the assembly of several subsystems : this is schematically shown in Fig. 1.

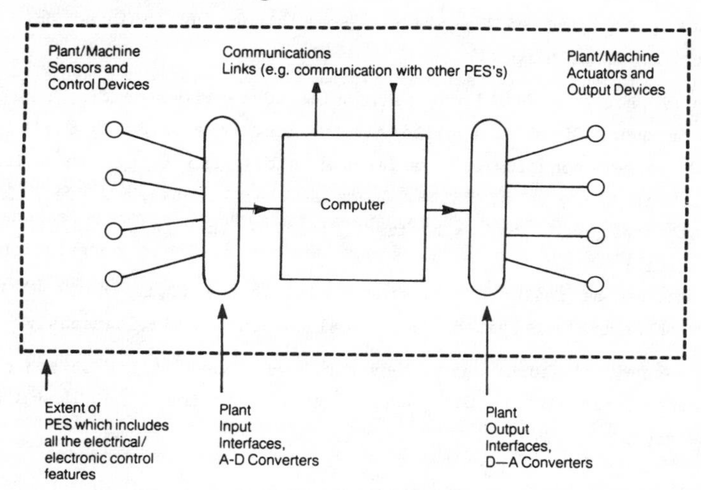

Fig. 1 : Schematical structure of a Programmable Electronic System (PES) (by courtesy of R. BELL, Health and Safety Executive, UK).

These subsystems are the following : the so called programmable electronics (PE), the input interfaces and output interfaces, the sensors, the actuators, and the communication links which can be of internal use (between subsystems) or external use (with specialized peripherals, or with other PESs). Each subsystem can be split up into several elements : for instance, the PE consists of a processing unit, of a memory unit, and of the joint power supply. The assembly of a PE with the associated I/O interfaces is generally labelled as programmable controller (PC).

When the various kinds of electromagnetic disturbances (either conducted, or radiated, or electrostatic discharges) are applied under various conditions (permanently or temporarily), the different subsystems as well as their components can exhibit a large variety of responses.

On one hand, responses to a given disturbance stimulus can differ functionally. For instance, a noise signal can be superimposed to the output signal delivered by a sensor although the sensor works normally. On the contrary, a storage area can be fully erased and then no longer functional.

On the other hand, the various subsystems can be EM susceptible in a selective manner : some can be rather affected by conducted disturbances (e.g. the processing unit), whereas some others by radiated ones (e.g. communication link, sensors).

Such problems should not be considered as PES-specific, since they are met in every field of applied electronics. However it is worth noting that the large technological variety of subsystems brings an additional complexity in the case of PES's. Among several aspects, two peculiar and important ones must be emphasized, because they are typical of EM disturbances :

- the common mode failures : a single disturbance (f.i. radiated) has an action on several subsystems or several components simultaneously ;
- the low-energy disturbances : they can have a very strong action on some components, especially solid-state microelectronics (VLSI), which work at low signal level.

SPECIFIC PES - SAFETY PROBLEMS WITH RESPECT TO EMC

In the presence of electromagnetic interferences, two fundamental differences exist between the PES subsystem(s) in charge of computation and the other subsystems. The first difference is due to the programmed - and hence sequential- mode of operation, and the second one to the memory function. Both of these differences, which result from the principle of PES itself, can be at the origin of specific problems.

Indeed, the functions operated by the processing circuitry of the programmable electronics change during the running sequence. Consequently the effects on the circuitry of a given EM interference can result in different kinds of failures with respect to safety. It depends on the function operational at the time of perturbation. The second point deals with a memory effect, as follows. The consequence of an interference can appear not immediately. Indeed its physical effect expresses as an erroneous control order or as an erroneous computed paramater, which is stored in the memory. The further use only of that stored data will trigger a safety failure. Another case can be imagined : the stored program itself is modified. Then the failure will appear further on during the same program run, or during the next one.

So, from the general comments above, the electromagnetic environment seems to be able to play an important complex role for PES safety. The experimental work now presented deals with the programmable controller : with respect to the other subsystems, its EM susceptibility takes most specifically the problem.

BEHAVIOUR OF INDUSTRIAL PROGRAMMABLE CONTROLLERS IN A SIMULATED INDUSTRIAL ENVIRONMENT

This Section summarizes results of an experimental work performed at the Institut National de Recherche et de Sécurité about EMI effects on the safety of PES subsystems, f.i. industrial programmable controllers (IPC). These IPCs, which are a kind of PE, are considered as components and sold as such. A measurement campaign was performed in 1983 on 37 IPC's (5), and a second campaign in 1985 on 45 IPC's (6). All together, 23 of the world biggest manufacturers were represented.

CHOICE OF TESTS

The tests have been performed at INRS in presence either of the manufacturer of the IPC, or of its distributor. A test set-up has been designed in order to experiment as close as possible to industrial conditions. The same set-up, schematically displayed in Fig. 2, was used for all the IPC's.

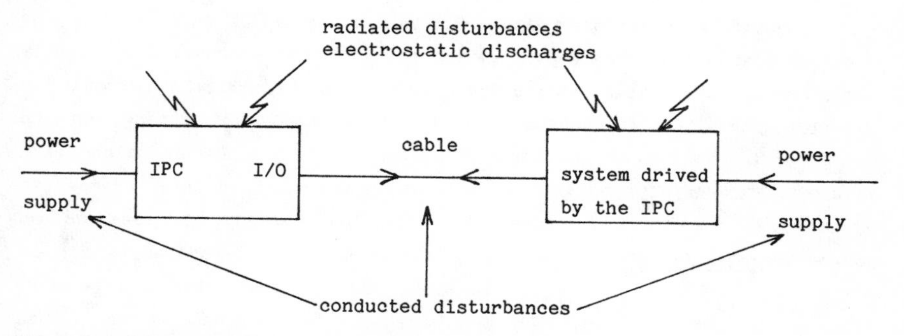

Fig. 2 : Schematical display of the test set-up.

The electrical disturbances are applied either on the power line of the IPC and of the drived system, or on the communication link between them. Electrostatic discharges are triggered on the cabinets.

The test procedure is as follows : the level of the parasitic signal is gradually incremented up to an upper limit, while the system behaviour and the user program run are monitored. When an abnormal behaviour occurs, the characteristics of parasitic signals are recorded.

Some differences exist between the tests actually performed and the tests stipulated by standards (e.g. NFC 63-850 (7) and IEC 801 (8)). On one hand, the purpose is different : only injurious effects of industrial disturbances are here of interest. No judgement on the quantitative level of performance is brought. On the other hand, the kind and the number of tets are not the same : a search for a correlation between various tests has been attemped. Unfortunately for the test simplification, no firm conclusion could be drawn on the design of an EMI type-test. All possible EM tests have to be completed.

PERFORMED TESTS

The list of the performed tests is given below. A detailed description has already been published (réf. to (5), (6) and (9)). The items are as follows :

- variation of power supply voltage (0 to 220 V) ;
- line failure (0 to 500 ms in duration) ;
- spike (few kV in common or normal mode) ;
- atmospheric transient overvoltage (up to 5 kV) ;
- electrostatic discharge (up to 21 kV) ;
- electromagnetic radiation (exploratory, frequency range : 20-250 MHz, E-field strength : few tens of V/m).

MAIN OBSERVED MALFUNCTIONS

Non-destructive interferences

- Deterioration of memories

The main occurrence happens in RAMs containing data or program instructions. The following changes are stated :

. change of a bit value within a word ;
. change of one or several words.

The consequences of such memory deterioration cannot all be described. Among them, one can mention :
. changes in the sequences of the program ;
. execution of a program not wished for ;
. stops during execution of a program, with unability to restart.

- Deterioration of output values

During some tests, one or several outputs of the PC are disturbed. The following behaviours have been stated :
. the outputs remain in the same state instead of switching ;
. the outputs switch ill-timely.

Of course malfunctions of the automated system occur, e.g. an ill-timed start of a motor, which can be the cause of injuries.

- Deterioration of timers

The timers are either numerical or analogical ones. Whatever it may be, the consequences of such abnormalities are as follows :
. the initial value of the timer changes for any other one ;
. the timer starts ill-timely again ;
. the output of the timer is no longer able to switch.

- Deterioration of the program counter

That kind of deterioration is close to the one stated for the memories. However, a refined analysis showed that sometimes, only the program counter was disturbed. Then the automated system executes a program which is not wished for (crippled execution, execution of another program, program-stop).

Remark :

Some automatized systems had a "watch-dog" function in order to detect malfunctions. In some cases the disturbed signals triggered the watch-dog, and the PC was set into a safe state. In some other ones, the watch-dog has not played its safety role. So this function alone cannot fix the whole safety of the system. With respect to the required safety level, the watch-dog, though necessary, should be used jointly with other protection methods (redundancy (10), dynamismus, selfmonitoring, signature analysis (11), and so on...).

Destructive interferences

The interferences have sometimes destroyed components such as fuse, varistor, resistor, capacitor, transistor, integrated circuit. Such incidents have especially been observed within the power supply unit. Furthermore, some modules in the central processing units or in the I/O devices have been hit too. The consequence is generally a program-stop with outputs locked either to zero or to the last operating state.

STATISTICAL RESULTS

Second campaign (1985)

About 30 % of the IPC's have completed the test set without malfunction. Table 1 and Table 2 summarize the results.

DISTURBED IPC's (PERCENTAGE)	EM INTERFERENCE
12	fast, low energy ($\sim$ 2 mJ), $\leqslant$ 2 500 V
5	normal mode, high energy ($\sim$2J), $\leqslant$ 1 000 V
28	common mode, high energy ($\sim$2J), $\leqslant$ 3 000 V
5	shock wave, $\leqslant$ 5 kV

Table 1 : Effects of EMI on IPC's.

COMPLIED IPC's (PERCENTAGE)	SEVERITY CLASS NR. AND UPPER VOLTAGE	
	fast transients (IEC 801.4)	
21	I	≤ 0.5 kV
21	II	≤ 1. kV
16	III	≤ 2. kV
42	IV	≤ 4. kV
	electrostatic discharges (IEC 801.2)	
17	I	≤ 2. kV
19	II	≤ 4. kV
17	III	≤ 8. kV
47	IV	≤ 15. kV

Table 2 : Distribution of IPC's versus severity class.

Comparison between 1st. and 2nd. campaigns

The results obtained from the same tests performed at two years of interval (1983 and 1985) are compared in Table 3. It is worth noting that about 70 % of the IPC's in the 2nd. campaign were new ones, i.e. not on the market in 1983).

MALFUNCTIONING IPC's (PERCENTAGES)		EM INTERFERENCES
1983	1985	
14	7	normal mode, low energy, ≤ 1.5 kV
47	28	common mode, high energy, ≤ 3. kV
15	5	shock voltage 5. kV

Table 3 : Comparison of IPC's susceptibilities.

Interpretation and comments

The statistical results displayed above show an improvement of the IPC behaviour with respect to EMI. It seems that manufacturers as well as users are aware of the problems arising from EM susceptibility.

However, the EM compatibility of PES's is of course necessary but not sufficient to get a good safety level. Indeed, the PES behaviour with respect to component failures must be analysed too, and with respect to all the programmation problems as well.

Electronic components are highly reliable when used in a protected environment. Actually in industry, the reliability can be much less. The protection against EM disturbances has allowed a lot of malfunctions to be suppressed, and hence to increase the lifetime of electronic units.

In order to solve the problem, manufacturers have been made sensitive : some positive results now appear. A few advices may be listed again for getting safety devices :

- to design according to state-of-the art ;
- to analyse the failures of components ;
- to perform some tests of the behaviour in a disturbed environment ;
- to check the manufacturing ;
- to know well the failures while the device is operating.

CONCLUSION AND PROSPECTS

In the paper above, the PES safety relationship with EMI has been approached from both conceptual and experimental points of view. The complexity of EMI-PES safety problems was emphasized on the bases of PES structure and of PES technology. Analogical electronics is there intricated with a programmable logical one. Both of these technologies can be EM susceptible, but likely nor in the same manner nor with the same effects. In addition, programmable sequential control can behave specifically when submitted to EMI. Experimentally, the results of EMI tests performed on world marketed programmable controllers show an important percentage still suffering of safety related malfunctions, in spite of recent improvements.

Now, several questions arise, to which prospective answers can be attempted.

i. Should EMI-PES safety problems be expected to grow ?
There are some arguments in favour of an affirmative answer. Electromagnetic pollution of the environment, especially the industrial one, spreads rapidly. It is a consequence of electromagnetic applications being widely developed for civilian, military, scientific, and of course, industrial purposes. The growing use of microelectronics in PES lets EM susceptibility increase because of the low level signals which are used.
At last, more and more sophisticated PES design, as observed, would probably be exposed to more and more sophisticated EMI failure modes.

ii. What are the EMI research topics to be investigated, in order to improve the related PES safety ?
They are of course numerous. A large part of them are relevant to EM susceptibility improvement. It ranges from studying the susceptibility of components, especially microelectronic ones up to designing PES architectures which minimize the problem. EMI coupling modes, modes of actions, failure modes, protections, and so on, should be studied.

As well as these topics, there is another fundamental one to be investigated, at a basic level : what are the actual EM disturbances which stimulate PES in the real industrial environment(s) ? To the end of designing at least representative tests procedures, the assessment of PES safety from the EMI point of view requires detailed answers to that question. Not much data seem to be presently available. Experimental investigations are to be carefully pursued in-situ. INRS is intending to bring in the future a contribution to such an objective.

REFERENCES

1 - GERMER, G., MEFFERT, K., Existing research deficiencies in the use of PES in industrial robot plants (in the present volume), 1986.

2 - DANIELS, B.K., Guideline Framework for the assessment of PES (in the present volume), 1986.

3 - BELL, R., Future Strategy for the use, development and review of the Guidelines Framework (in the present volume), 1986.

4 - VAUTRIN, J.P., DEI-SVALDI, D., Installations robotisées en France : conséquences sur l'hygiène, la sécurité et les conditions de travail. Résultats de l'enquête INRS-CRAM, Cahiers de Notes Documentaires, 120, 351-359, 1985 (translation in preparation, to be published by Elsevier, 1986).

5 - DEI-SVALDI, D., VAUTRIN, J.P., Comportement des automates programmables en ambiance industrielle, Electronique Industrielle, 64, 29-34, 1984.

6 - DEI-SVALDI, D., Electronique Industrielle, to be published, 1986.

7 - Norme française homologuée NFC 63-850 (octobre 1982), Automates programmables, UTE, 12, place des Etats-Unis, 75783 PARIS CEDEX 16.

8 - Norme CEI n° 801-1/2/3, 1985, Compatibilité électromagnétique pour les matériels de mesure et de commande dans les processeurs industriels, CEI, 12, place des Etats-Unis, 75783 PARIS CEDEX 16.

9 - DEI-SVALDI, D., VAUTRIN, J.P., Les automates programmables : nouvelles technologies, nouveaux risques, principes de sécurité à appliquer, Cahiers de Notes Documentaires 117, 467-473, 1984, INRS, PARIS.

10 - KOPKA, B., GERARDIN, J.Ph., Amélioration du niveau de sécurité des systèmes à microprocesseur par application du concept de redondance, Electronique Technique et Industries, 24, 43-54, 1985 (Part 1) and 25, 47-54, 185 (Part 2).

11 - SCHWEITZER, A., Improving the safety level of Programmable Electronic Systems by applying the concept of signature analysis (in the present volume), 1986.

IMPROVING THE SAFETY LEVEL OF PROGRAMMABLE ELECTRONIC SYSTEMS (PES) BY APPLYING THE CONCEPT OF SIGNATURE ANALYSIS

Alexis SCHWEITZER

INSTITUT NATIONAL DE RECHERCHE ET DE SECURITE (INRS)
Service "Electronique - Sécurité des Systèmes"
54501 VANDOEUVRE CEDEX - FRANCE

SUMMARY

The concept of signature analysis is used here on-line differently from the conventional on-maintenance use. Its use should make it possible to achieve a very high capacity of fault detection. The advantage of a check during operation of the system is to be able to monitor in real time the validity of task sequencing carried out by the application software, in order to obtain systems with a high level of positive safety (fail safety).

This contribution describes in detail the principal functions and characteristics of a signature analyser module. It is designed to measure virtually all the numerical information available on the data bus of several types of microprocessor.

The general methodology is then presented for the optimum use of the module : based on a study of existing software structures, it assumes the use of a structured modular programming technique. The solutions given use a simple decomposition of the software into several blocks in which the information to be analysed is determined.

Finally, the results obtained are given. The safety level is assessed by simulating physical and software faults using an automatic tester developed by INRS. The behaviour of the device when submitted to electromagnetic disturbances by conduction and radiation is also planned.

INTRODUCTION

The use of VLSI processors and components is rapidly developing in industry and provides many advantages as regards reliability, cost savings and variety of systems. But against this, problems inherent in failure modes of components and in electromagnetic interferences arise at the safety level. This is most worrying for the control logics of dangerous systems (automatised systems, manipulator robots, dangerous machines,...) or the staff protective devices (light guards, sensitive mats and floors...). As a matter of fact, no reduction in the safety level can be allowed between a conventional- and a programmed-type designing.

There are general solutions for compensating these drawbacks : redundancy, item dynamism, self-monitoring, watch-dog...). Where properly applied they enable satisfactory safety levels to be achieved. However, through the use of a microprocessor and its associated software, new methods can be developed. This is the case for signature analysis which is a powerful detection and fault diagnosis tool, well adapted to digital techniques. The HEWLETT-PACKARD company has promoted this method. It is presently used in maintenance for localising faulty components and in production for checking the quality of digital cards at the end of the production lines.

The aim of the investigations presented here is to propose an adaptation of the signature analysis concept to systems safety, which is original, cost saving and as universal as possible but for which the improvement of the safety level obtained could be globally assessed. The method proposed must be considered as an on-line tool working in real-time, based upon signature analysis.

First, we introduce the concept of signature analysis (see section "THE CONCEPT OF SIGNATURE ANALYSIS"), its adaptation to safety (see

"PRINCIPLE OF IMPLEMENTATION OF SIGNATURE ANALYSIS"), then the example of a real case with the results obtained (see "RESULTS OBTAINED IN A REAL CASE") and lastly, the possibilities to be expected from this method (see "CONCLUSION").

THE CONCEPT OF SIGNATURE ANALYSIS

PRINCIPLE (ref. to (1))

The two basic elements of signature analysis are the use of a stimulus generated by the board under test and the compression of data sent by the point to be tested. For a card with a microprocessor, additional software for signature analysis should be planned at the designing stage. Its role will consist in periodically stimulating the whole of the card components. For a maintenance application (see the example in Figure 1), the serviceman should isolate the faulty component by measuring the signatures at different points of the card according to a procedure given by the designer. The faulty component is the one which, the most upstream, delivers a signature different from the one expected.

METHOD OF MEASUREMENT

Measurements are performed using a feedback shift register taking in to account, on one hand, the digital data stream produced by the chip under test and on the other hand, the controls determining the duration and the synchronization of the measurement. This technique refers to CRC's coding (cyclic redondancy control) used in serial transmissions. Its major characteristic is to enable pseudo-random sequences to be obtained. This means that if the input is at a constant zero logic level, at time t, the feedback shift register contents are equal to those at time t + N cycles, N being the reproducibility period for a shift register of n bits ($N = f(n) \simeq 2^n - 1$). For n = 16, N = 65535. It must also be noted that for differentiating a signature from an hexadecimal number during the readout on the existing measurement devices, the hexadecimal values from A to F are coded A, C, F, H, P, U respectively, in order to avoid any ambiguity.

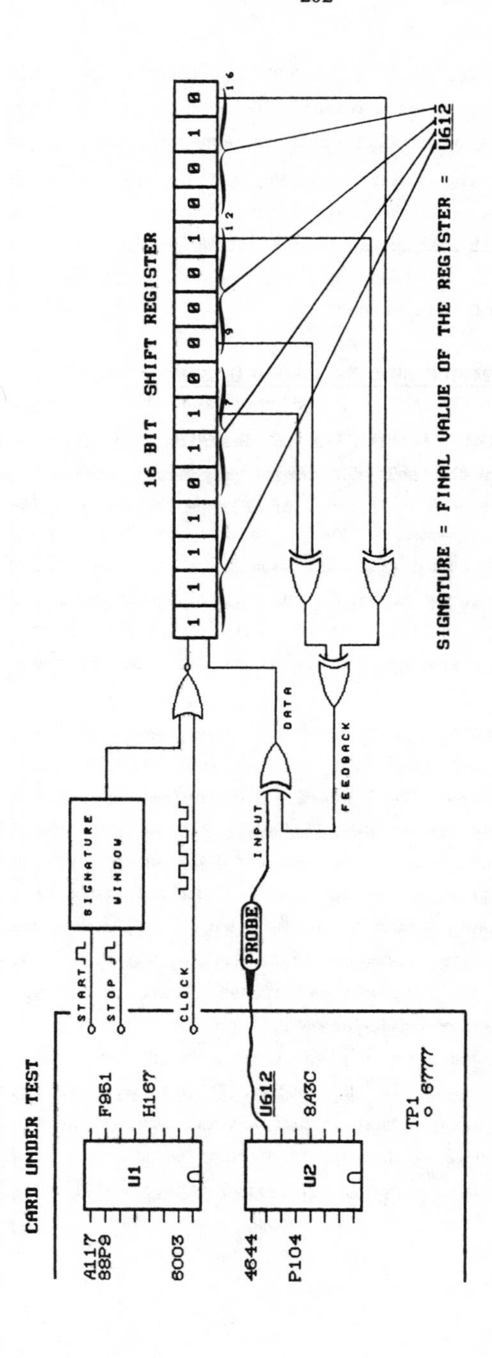

FIGURE 1 : PRINCIPLE OF SIGNATURE ANALYSIS IN MAINTENANCE

EFFICIENCY OF THE METHOD

The second major characteristic of signature analysis is its probability of detecting faults (or reliability F) which is equal to 99.998 % for a 16 bit shift register ($F \simeq 1 - 2^{-n}$ in %). In other words, the probability of obtaining an identical signature from a different digital data stream is in the order of 0.00002. This shows clearly the detection capability of this method.

PRINCIPLE OF IMPLEMENTATION OF SIGNATURE ANALYSIS

The aim of this section is to describe the different hardware functions existing on the self-monitoring signature analysis unit. These are shown in the schematic diagram of Figure 2. They have two major functions :

- the acquisition of the reference signatures ;
- the real time checking of the signatures measured on the system.

OPERATING PRINCIPLES OF THE UNIT

For making a real time unit, the main idea consists in simultaneously operating the unit and the system. The application software in the program-memory determines the overall operation of the system. The solution chosen is the use of an additional memory concurrently with the program-memory. This "signature analysis" memory has all the functions necessary for the operation of the unit. These are defined by the user together with the application software. Next section explains how to program this memory for the two operating modes of the unit.

PRINCIPLE OF THE SIGNATURE MEASUREMENT

The method of measurement is, of course, very similar to that quoted in the previous section, at least as regards its principle. However, there are some differences. The first one is the use of an 8 bit shift register. So, measurement is less accurate and the probability of detection is reduced to 99.6 %. The second concerns the processing of data

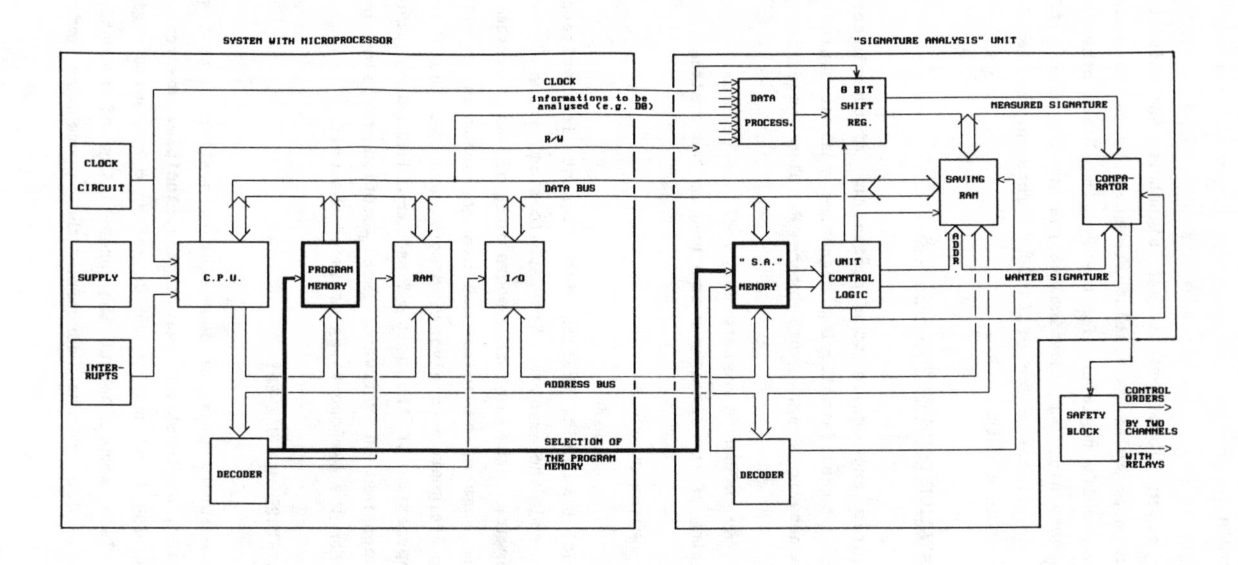

Fig. 2 : SCHEMA OF THE SELF-MONITORING UNIT BY SIGNATURE ANALYSIS to be analyzed. As a matter of fact, an analysis of several signals in a cyclic or a parallel way in real time can be envisaged. These means enable the lack of accuracy at the level of the measurement to be offset. The choice of the data to be analyzed is also an essential criterion to take into account. For the systems with microprocessors, the data transferred onto the data bus are very representative of the functioning of the system and of the internal transfers performed. Thus, they must be exploited first and foremost.

PRINCIPLE OF SIGNATURE ACQUISITION

Signature acquisition requires the progamming of the "signature analysis" memory in the "acquisition" mode. When the application software has been executed, the module measures signatures that it saves in a random access memory (RAM). The user can read these signatures when the system has stopped. It is better to renew this procedure several times, in order to check that the signatures obtained are repetitive. If this is the case, these signatures should be considered as a reference.

PRINCIPLE OF SELF-MONITORING IN REAL TIME

It is also necessary to program the "signature analysis" memory in the "self-monitoring" mode, just as it is for the "acquisition" mode. Actually, few changes are made between the two operating modes of the unit, except for some functions to be replaced by others. While the system is in operation the "comparison" function should be periodically activated at the end of each signature measurement. Then, it is easy to check the agreement between the measured signatures and the reference signatures by observing the output of the comparator at the commutation times.

ROLE OF THE SAFETY BLOCK

The safety block is not really a part of the unit because the action of the safety positioning depends on the system considered. Certain systems have a safety state which must immediately be reached in case of fault. Others require a special procedure prior to the complete stop, and this can be achieved only with a supplementary system in perfect working order. In any case the role of the safety block is to control a reliable safety positioning of the system. Thus, a compromise between reliability and the desired response time has to be studied.

In the present case, processing by the safety block concerns a pulse signal. The safety positioning is activated if the absence of pulse has a duration exceeding the maximal time allowed. This type of processing is necessary but not sufficient to provide a safety positioning whatever the fault considered. A possible solution consists in adding a software check to the hardware check. Its function is to check the sequencing of the signatures measured by the "Signature Analysis" unit.

METHODOLOGY

In order to apply in an optimal way the concept of signature analysis, a complete procedure has been adopted. The procedure to be followed and the solutions proposed are valid if the following conditions are adhered to :

- utilization of a structured modular programming technique ;
- writing of the application software into "assembler" language.

An overall analysis is performed from the flow-chart of the application software. The software is then broken down into a certain number of blocks made of a combination of the following structures :

- sequential structure ;
- repetitive structure or loop ;
- alternative structure.

Each block is then thoroughly analysed from the application software listing written in "assembler" language. A solution for the implementation of the signature analysis specific to each block should be chosen according to the following criteria :

- types of internal variables ;
- types of structures ;
- running time of each block.

RESULTS OBTAINED IN A REAL CASE

DESCRIPTION OF THE DEVICE

The method presented is applied on a device simulating the functioning of a light guard. This simulation is achieved by a programmed logic using the MOTOROLA 6809 microprocessor. Four infrared emitters are cyclically activated every 4 ms one after another.

IMPLEMENTATION OF SIGNATURE ANALYSIS

The application software is made of a permanently iterative structure. One of the four emitters is activated every time. An internal variable in the loop indicates the transmission in process. The implemented solution of the signature analysis chosen consists in the analysis of line D0 of the microprocessor at each run through the loop. Thus, the device is characterized by four different signatures, since there is a signature checking every millisecond. The self-monitoring unit, and more precisely the safety block, has been designed so that an error can be detected in 1.5 ms. This detection time is actually adjustable and depends on the application chosen.

FAULT-HYPOTHESES STUDIED

In order to assess the performances of the self-monitoring unit, the light guard has been submitted to a test procedure using the "DEFI" apparatus (2) which has been designed at INRS. It enables us to introduce faults into the central unit of programmable electronic systems and to observe the consequences induced in their functioning. The faults simulated by this apparatus range from mere stuck-at faults to multiple internal faults in the microprocessor, whether temporary or permanent. This test is not exhaustive but is very stringent. Thus, the safety level obtained can be satisfactorily assessed. In this case, we can speak of a coverage rate Tc of a system which is equal to :

$$Tc = \frac{Ndd}{Ndi} \text{ in \% where :}$$

Ndd = number of faults detected

and Ndi = number of faults introduced

BEHAVIOUR OF A DISTURBED SYSTEM AND RESULTS OBTAINED

Out of 17085 faults introduced 16473 were detected using a signature analysis check, hence Tc = 96.4 %. For the 612 faults not detected, a detailed analysis of the behaviour of the device was performed and gave the following results :

- the functioning of the device was identical to correct functioning in 1.6 % of cases ;
- the functioning of the device was slightly damaged but fulfilled its task in 0.2 % of cases ;
- the functioning of the device was severely damaged but fulfilled its task in 1.4 % of cases ;
- the functioning of the device was severely damaged and not fulfilled its task in only 0.4 % of cases.

Only the last-mentioned situation is really dangerous. This means that only a very low percentage of faults (0.4 %) has not been detected by the self-monitoring unit and induced abnormal behaviour in the system. This result is to parallel with that of the theoretical probability of detecting faults by signature analysis which is of the order of 99.6 % in the present case. Even if these results are particular to a specific device, they are very promising for the future of signature analysis.

CONCLUSION

Some of the numerous possibilities offered by signature analysis and implemented in the unit proposed are described in the previous sections. The unit is entirely independent and almost all-purpose because it is linked to the system by a set of lines existing on most microprocessors (data and address buses, clock, program-memory selection and reading/writing signals). Moreover, it is equipped with multiple powerful and evolutive functions, which lead us to expect varied and interesting possibilities of use. The results obtained confirm these possibilities too. The prime aim of the unit is to check the proper functioning of programmable electronic systems during service, in order to ensure a safety task. The results to be expected from the use of signature analysis depend on the extent to which the method is used because in certain cases, it should be applied only on a part of the application software. It also depends on the quality of the information analysed. This depends on the complexity of the systems considered and on the structures of its application software. Furthermore, the existing unit could be adapted for realizing a self-diagnosis function. Thus it could become a tool of use in other fields such as availability and maintainability.

Finally we would like to add that the work presented has been performed within the framework of a Ph. D. thesis which will be presented-defended in 1986.

REFERENCES

1 - Application Note 222-2, Application articles on signature analysis, HEWLETT-PACKARD Publication 02-5952-7542, May 1979.

2 - GERARDIN, J.Ph., Notice d'utilisation de l'appareil simulateur de défauts de l' I.N.R.S., INRS Service "Electronique - Sécurité des Systèmes", VANDOEUVRE, FRANCE, mai 1985.

THE PHYSICAL SIMULATION OF FAULT: A TOOL FOR THE EVALUATION OF PROGRAMMABLE CONTROLLERS (PC's) BEHAVIOUR ON INTERNAL FAILURE

J. L. TRASSAERT

Automobiles PEUGEOT
On behalf of the French Electrotechnical Committee
(UTE/CEF 65/GE1)

INTRODUCTION

The evolution of the volume of automation in all the branches of the industrial activity is related with other aspects, particularly

- an increasing complexity of the problems to be solved
- an extension of the field covered by the automation equipments
- on imbrication of systems.

The industrial requirements are concurrently evolving. Master words are

- Zero failure : Reliability and availability in order to guarantee the continuity of the production,
- Zero fault : quality of the product,
- Safety of operation in order to guarantee the safety of persons and goods.

All these requirements are parts of the task of the automatician who has to comply with the DRASTIC REQUIREMENT:

zero fault + zero failure + safe operation

To satisfy this demand, the best tool is the Industrial Programmable Controller, with both its qualities and defects.

The French Standard NF C 63 850 (1982)

This Standard determines the technical and constructional characteristics together with the tests to be applied to the Programmable Controller (PC) in specified conditions of use.

A French Working Group (UTE/CEF 65-GE1) has been set up in Autumn 1983, in order to complement for the French Standard and to specify a method for the evaluation of the PC behaviour in the occurrence of a fault and to define general classes of use, according to this behaviour.

This group aims to:

- conclude by the end of 1986
- present the result of its work as a contribution to international standardisation.

The type of behaviour on internal fault

The French Standard lists 5 possible types of behaviour:

Type 1 : The fault is detected, reported and the PC continues to operate,

Type 2 : The fault is not reported and the PC continues to operate. Such a failure can only be revealed by a routine check specified by the manufacturer,

Type 3 : The fault is detected and signalled, the outputs change into a fixed state,

Type 4 : The fault is detected and signalled, the outputs change into a not fixed state,

Type 5 : The fault is not reported, the outputs change into a not fixed state.

A Sixth behaviour is to be considered which is not in the French Standard, but may be observed when testing.

Type 6 : The fault is not reported, the outputs change into a fixed state.

As a conclusion: the consideration of two criteria defines the types of behaviour:

- first criterium : detection and signalisation on one hand
- second criterium : behaviour of the output on the other hand.

The criterium "detection and signalisation" is binary.

The criterium "behaviour of the outputs" is ternary:

1 - the PC continues to operate
2 - the outputs change into a fixed state on internal fault
3 - the outputs change into a not fixed state on internal fault.

The requirements of the user

Depending on the application, the user may express requirements of different natures on internal fault:

- signalisation for the operator (visual indicator) and for the other components of the automatism,

- fall back position for the outputs, e.g: each one at zero (a state which looks like being the most commonly chosen),
- availability of the system maintained. This requirement is generally obtained by the choice of adequate systems building based on the knowledge of a fall back position for the PC outputs (redundant equipment) from a fall back position of the PC. It does not exclude the designing of PCIS with a high availability degree.

Methods of evaluating of the behaviour on internal fault

Several ways have been investigated:

1 - software simulation
2 - analysis on a schematic diagram
3 - Physical Simulation of Fault (PSF).

Software simulation is very complicated even for relatively simple diagrams. It involves a scrupulously updated modelisation of components and heavy informatical tools.

Analysis on schematic diagram quickly proves to be tedious and often incomplete. It does not always take into account the operational hazards generated by hardware or software faults.

Experience shows that Physical Simulation of Fault (PSF) is the most relevant and objective way. As a matter of fact, due to its systematic aspect, it enables it to examine a great number of cases and to give rise to actual behaviour through the simulation of faults considered as representative, and which can occur:

- either directly,
- or by analogy on the level of consequences.

For these reasons, this method has finally been adopted and investigated. It is the one that we intend to propose here.

Physical Simulation of fault

1 - unplugging the modules
2 - unplugging cards
3 - fault on connections
4 - withdrawal of integrated circuits (IC)
5 - simulation of stuck at fault on pins of ic integrated circuits (see details hereafter)
6 - simulation of an alteration, definitive or temporary, of the control sequence.

The simulation of the first 4 types does not require any special tool.

The simulation of alteration of a control sequence corresponds to a fault resulting from the failure of a component, or from a disturbance in the central processing unit (CPU). This type of simulation is subject to studies and experiments of INRS (French National Institute for Safety). The method of simulation is described in Reference 4 of the bibliography at the end of this paper.

For the simulation of failure on the pins of IC, it is necessary to have testing possibilities as described below.

Construction of a simulator

Principle

- the simulator has to be inserted between the IC and its support
- each pin is galvanically isolated
- a signal simulating the consequence of a fault is substituted to the real signal,
  - stuck at low or high state
  - high impedance state
  - oriented short circuit.

Example: For the output of an integrated circuit. An output may be at low or high state. The diagram is then as follows:

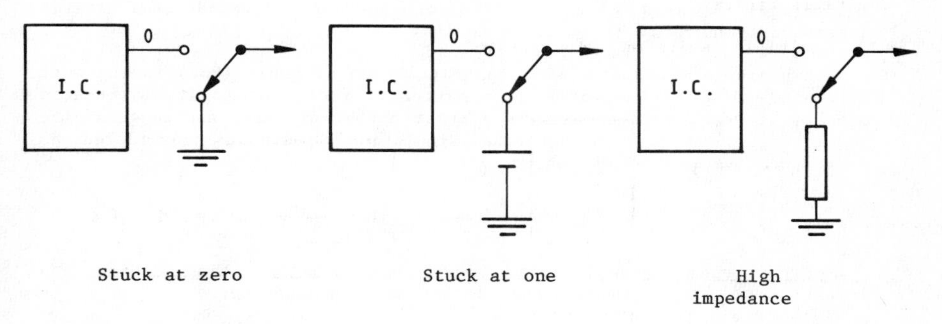

Stuck at zero

Stuck at one

High impedance

Orientated short circuits

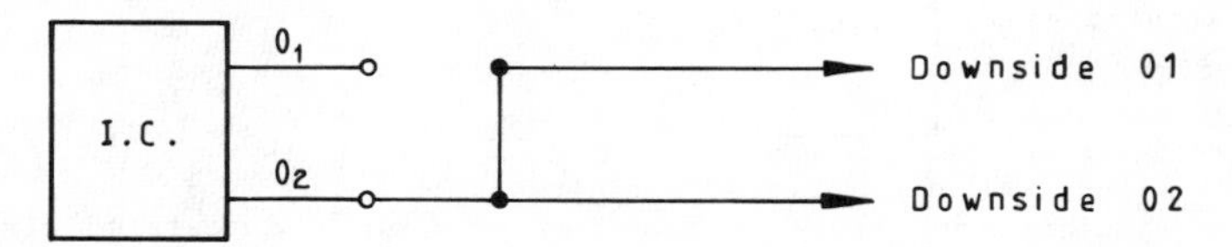

Orientated short circuit 02 towards 01

The signal of 02 is substituted to the one of 01.

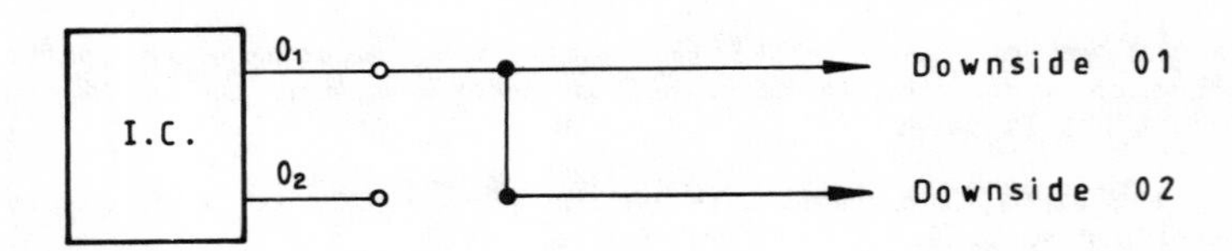

Oriented short circuit 01 towards 02

The signal of 01 is substituted to the one of 02.

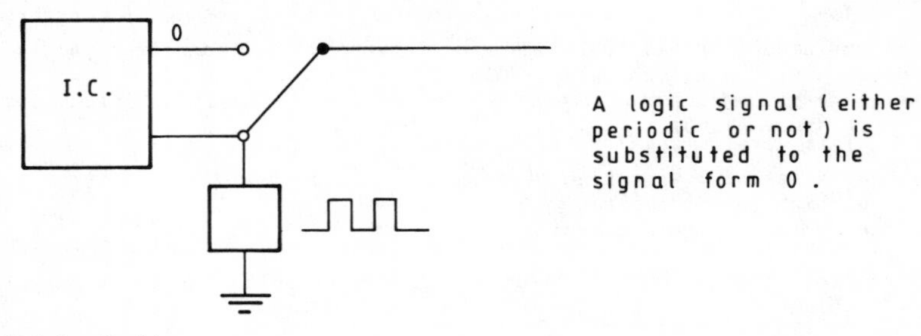

Multiple faults

Multiple faults may be simulated

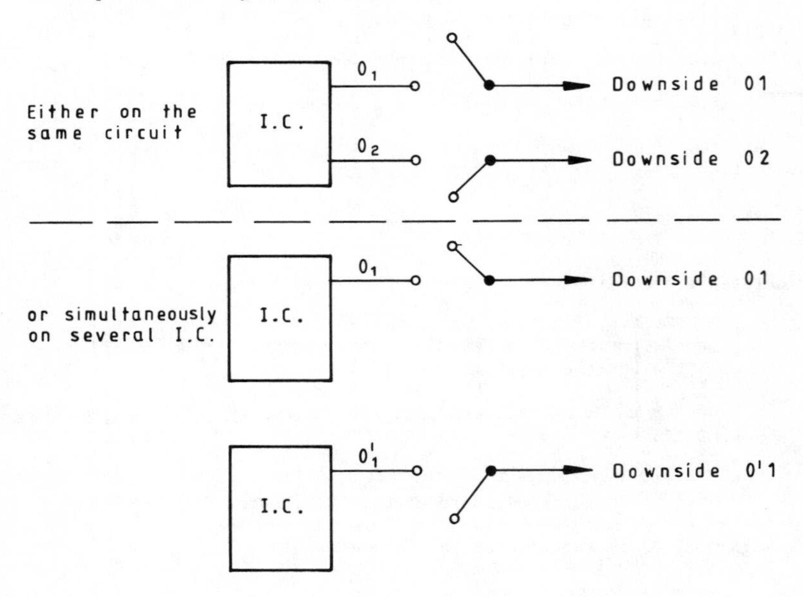

Technological choices

1 - Manual version (I)

With a panel of switches it is possible to simulate:

- stuck at low or high state
- oriented faults
- multiple faults.

Such a simulator allows 30 or 40 faults to be simulated every day. The great number of manipulations is a cause of errors. The intermittent faults are not simulated.

2 - Automatic version with relays - (II)

This version uses electromagnetic micro relays.

The size is rather bulky.

Multiple faults and intermittent faults are not simulated. The duration of commutation exceeds 1 ms.

3 - Automatic version with semi-conductors VMOS - III

Small dimensions.

Speed allowing intermittent faults to be simulated (10 MHz).

One hundred faults may be simulated a day.

In laboratory, the three versions are used.

As the description of these mounts have been subject of other publications (see Appendix), they are not described in detail.

Experience gained from the tests

Testing the method made it necessary to define:

- a minimum physical configuration, consisting of the CPU, the user's memory, one digital input module, one digital output module and the supply,

- an application software when it is needed. It is simple and described by a flow chart or one GRAFCET.

The CPU and Memory cards use a number of integrated circuits greater than the interface cards (I/O) where many discrete components are not implicated in the method of simulation. As a result:

- the method is particularly relevant for the cards CPU and Memory cards,

- for the interface cards, the method of physical simulation of the fault has to be imperatively completed by a diagram analysis.

The analysis is generally simple: it concerns the passive components the fault modes are known of which and not numerous (two or three at the most).

The tests have shown a basically different behaviour for the cards of the CPU (and Memory) and for the cards of the I/O's interfaces:

- On the CPU, "sleeping" failures appear that is to say, faults which are not signalled and cause not dysfunction.

  The number of sleeping failures may reach up to 20% of the simulated faults on the CPU.

An accurate analysis of this kind of fault leads to differenciate them according to their consequences.

- deterioration of the reliability in the time if the fault persist,
- danger for the process if a second fault occurs,
- they do not directly affect the process but disturb the maintenance.

- On the I/O's cards, there appear either some blocking up or wrong orders.

Tables and diagrams below show some example of the results obtained by the application of the method on some models of PC's.

Table resuming the results for each card

TABLE 1

CARDS	CPU						MEMORY					
Type behaviours	1	2	3	4	5	6	1	2	3	4	5	6
Withdrawal of card			100						100			
Fault of connection	0	47	43	2	2	6	0	38	56	0	3	3
Withdrawal IC	0	7	7	0	7	78	0	40	80	0	0	0
Fault on pins	1	19	7	0	2	70	0	21	55	0	10	12

TABLE 1 (cont'd)

CARDS	OUTPUT INTERFACE						INPUT INTERFACE					
Type behaviours	1	2	3	4	5	6	1	2	3	4	5	6
Withdrawal of card			100						100			
Fault of connection	0	14	35	35	7	7	7	27	60	0	7	0
Withdrawal IC	0	0	100	0	0	0	0	0	60	0	40	0
Fault on pins	0	0	58	30	9	3	0	3	41	22	26	6

GLOBAL RESULTS FOR THE PHYSICAL CONFIGURATION
(HERE PCU AND MEMORY)
AND THE I/O INTERFACES FOR THE PROGRAMMABLE CONTROLLER

TABLE 2

CARDS	PHYSICAL CONFIGURATION						INPUT/OUTPUT					
Type behaviours	1	2	3	4	5	6	1	2	3	4	5	6
Withdrawal of cards			100						100			
Fault of connection	0	43	49	1	2	4	3	21	48	17	7	3
Withdrawal IC	0	14	19	0	6	61	0	0	79	0	21	0
Fault on pins	1	20	19	0	4	56	0	2	50	26	18	5

CARDS	PROGRAMMABLE CONTROLLERS					
Type behaviour	1	2	3	4	5	6
Withdrawal cards			100			
Fault of connection	1	38	49	5	3	4
Withdrawal IC	0	10	36	0	10	44
Fault on pins	1	14	28	8	8	42

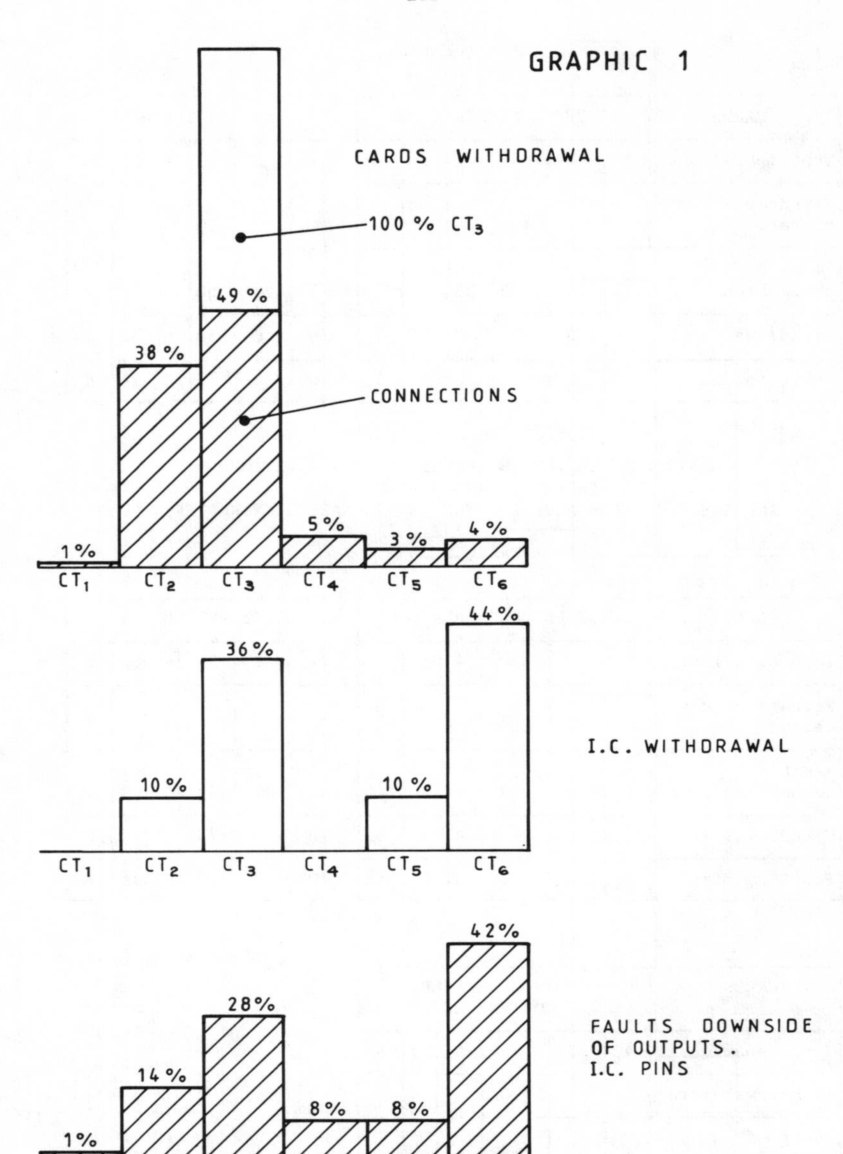
GRAPHIC 1
CARDS WITHDRAWAL
100 % CT3
49 %
38 %
CONNECTIONS
1 %
5 %
3 %
4 %
CT1
CT2
CT3
CT4
CT5
CT6
44 %
36 %
I.C. WITHDRAWAL
10 %
10 %
CT1
CT2
CT3
CT4
CT5
CT6
42%
28%
FAULTS DOWNSIDE
OF OUTPUTS.
I.C. PINS
14 %
8 %
8 %
1%

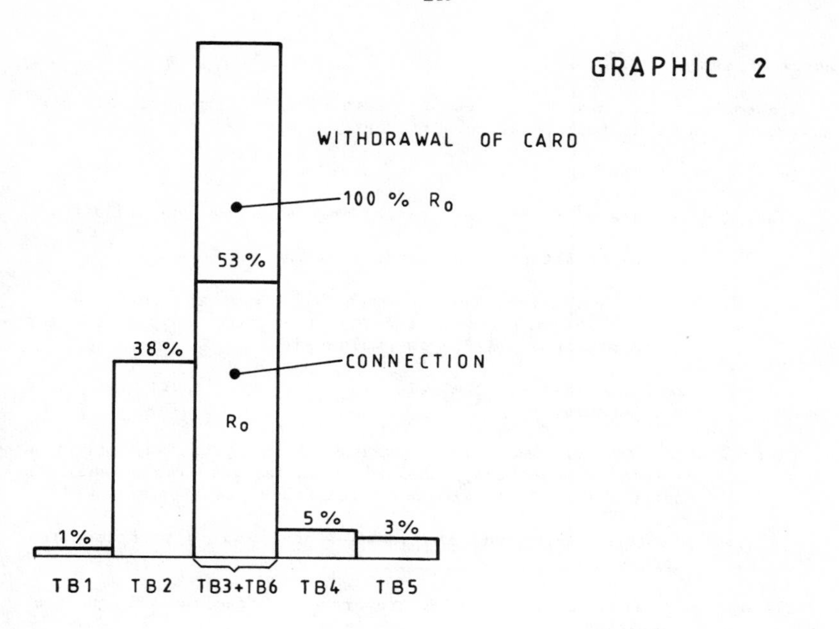
WITHDRAWAL OF CARD
100 % R_0
53 %
38 %
CONNECTION
R_0
1%
5 %
3 %
TB1
TB2
TB3+TB6
TB4
TB5

GRAPHIC 2

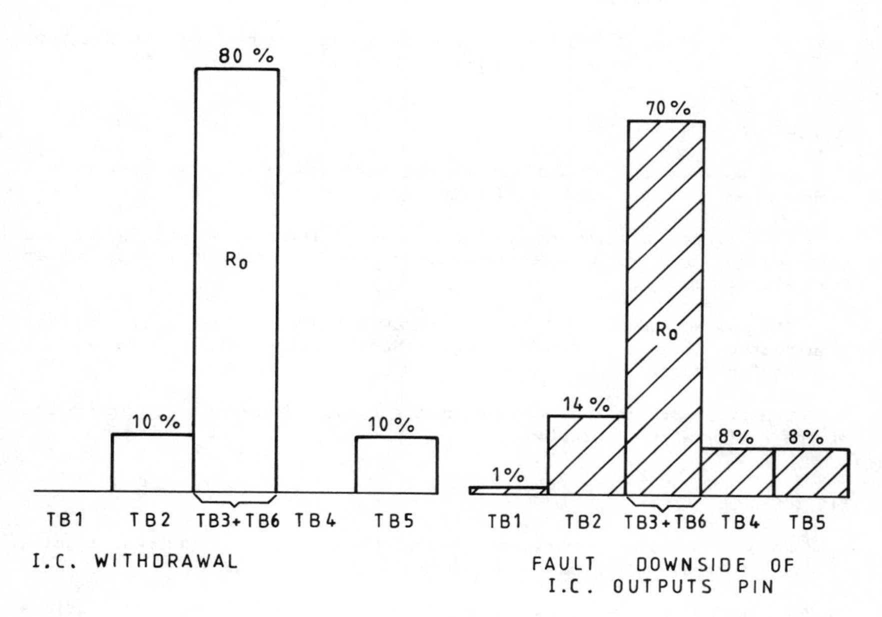
80 %
R_0
10 %
10 %
TB1
TB2
TB3+TB6
TB4
TB5
I.C. WITHDRAWAL
70 %
R_0
14 %
8%
8%
1%
TB1
TB2
TB3+TB6
TB4
TB5
FAULT DOWNSIDE OF
I.C. OUTPUTS PIN

Classes of use

According to the use of the PC's, the requirements, regarding their behaviour on internal fault, may be different.

It appears that three classes may be considered.

Class 1 : in case of fault (here established with main supply off).

- the PC detects and signals the fault,
- the efficiency (or coverage factor) of the software for testing when switching on the main supply or for maintenance helping is evaluated,

In this class, the aim is only to report and help maintenance.

Class 2 : In case of fault (here simulated on operation), after a specified duration, the PC signals the fault and its outputs turn to a zero state fall back position zero.

In this class, the aim is to signal and to position at a zero state.

Class 3 : In this class, the PC reports the faults and goes on operating.

The aim of this class is a high degree of availability. This class is under consideration.

CONCLUSION

The work of the WG UTE/CEF 65 GE1 has permitted us to elaborate a method which is based on effective tests.

Some French manufacturers who take part in this working group use this method to check the performances of their equipment at the better designing stage.

The main users of these equipments can now classify their requirements in their specifications. Such are the energy producers and car manufacturers.

Thus, the French WG makes an important contribution to the evaluation of the PES's safety operation.

BIBLIOGRAPHY

1. Norme francaise homologuee: Appareillage industriel a basse tension - Automates programmables NF C 63-850 (Octobre 1982).

2. F. MORILLON : Methodes et outils de tests de cartes - Simulations physique de defauts.

2eme Colloque Interntional sur la faibilite et la maintenabilite PERROS GUIREC (Septembre 1980).

3. F. MORILLON : Automates programmables. Securite et comportement en presence de panne interne.

4eme Colloque International sur la fiabilite et la maintenabilite PERROS GUIREC (Mai 1984).

4. J.Ph. GERARDIN : "DEFI" (Defined faults injection) - electronique industrielle (novembre 1984).

Members of the WG UTE/CEF 65-GE1

Manufacturers and Manufacturers association

MM	RAGE	CGEE ASLTHOM
	CUNZI	CROUZET
	LAMOTHE	TELEMECANIQUE
	ANTOINE	GIMELEC

Users and users associations

MM	MORILLON	EDF
	CHEMALI	EDF
	PIAT	EXERA
	TRASSAERT	CNOMO

Organisation of Agreement

MM	CORRIHONS	BUREAU VERITAS
	GERARDIN	INRS

GIMELEC - GROUPEMENT DES INDUSTRIES DE MATERIEL D'EQUIPEMENTS ELECTRIQUE ET DE L'ELECTRONIQUE INDUSTRIELLE ASSOCIEE.

EDF - ELECTRICITE DE FRANCE

CNOM - COMITE DE NORMALISATION DES OUTILLAGES ET MACHINES-OUTILS.

EXERA - ASSOCIATION DES EXPLOITANTS D'EQUIPEMENTS DE MESURE DE REGULATION ET D'AUTOMATISME

INRS - INSTITUT NATIONAL DE RECHERCHE ET DE SECURITE.

SAFETY WITH NUMERICALLY CONTROLLED MACHINE TOOLS

SUNDQUIST, MATTI

The National Board of Labour Protection
P.O. Box 536, 33101 TAMPERE
FINLAND

ABSTRACT

The study deals with the safety with numerically controlled machine tools. In addition to the technical devices the organization of numerically controlled machining is also discussed. Data were collected mainly on experiences gained from the operation of and occupational safety with numerically controlled machine tools in factories and workshops. Causes for personal accidents and material damage came up in all phases of the numerically controlled industrial process.

A model of the risk factors with numerically controlled machine tools has been developed using the Systems Approach, and the critical points in the numerically controlled machining process have been defined and remedies have been sought.

SIGNIFICANCE OF NUMERICALLY CONTROLLED MACHINE TOOLS

The use of numerically controlled machine tools became significant in Finland in the 70's, when the number of machines increased from a few dozen to over a thousand. Today, the share of numerically controlled machine tools is about one fourth of all the machine tools sold in the country. This means that presently their total number amounts over to two thousand. Thus the significance of numerically controlled machine tools for engineering industry and its working conditions is far greater than that of industrial robots, whose influence on working life is under lively public discussion.

The impact of numerically controlled machine tools on occupational safety is similar to that of automatic machines in general: the total number of occupational accidents decreases, as the worker can be protected from dangerous processes but on the other hand, the potentiality of serious accidents increases due to new and unpredictable hazards. In Finland there have been two fatal industrial accidents in which numerical control has played an important part. In both cases machine parts moving by engine power collided with each other and the flying debris hit the worker. In the first case, a chip got into the control panel and caused a change in the program with the result that the tool-carrying slide made a faulty movement. In the second case, the worker forgot to skip a so called block delete, whereby the tool-carrying slide collided with the chuck. A serious accident happened when the cutting tool collided with the chuck in connection with programmed tool changing. There have been several similar near accidents, but usually the flying debris have remained inside the machine guards and so injuries to persons have been avoided. Material damages may, however, have amounted to hundreds of thousands of marks and the stoppage time to several months.

These cases have given a cause for a closer study of accidents with numerically controlled machine tools (Sundquist, 1983). Experience shows that the introduction of automatic machines does not as such improve working conditions or occupational safety. Whenever new technology is introduced, all problems cannot be foreseen or solved, and unexpected situations arise, which may involve possibilities of accidents. Trial and error does not yield favourable results from the very beginning.

HOW THE STUDY WAS CARRIED OUT?

A study aiming at improved occupational safety usually starts with collecting data related to the subject and with a risk analysis of the object of study. Experiences from the use and safety of numerically controlled machine tools were gathered by inquiries to authorities and research institutes in different countries and by an inquiry in Finland into about 30 enterprises using numerically controlled machine tools. In addition, interviews and observations were made in some ten enterprises. Experiences from use were still rare in the 70's, so statistical processing of the material was not possible. The study was concentrated on a qualitative analysis, that is finding out the so called critical points in numerically controlled machining processes. This means that the whole working process is described starting from designing the workpiece and proceeding to working out the machining program and then through the machine setup to the actual machining. All the work phases related to machining are specified as to the working methods and the technical equipment, and those errors that can lead to hazardous situations are charted out (Cf. Fig. 1).

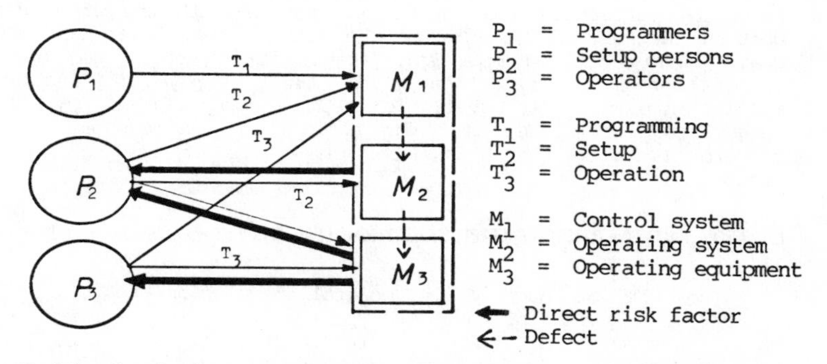

Fig. 1 Model of relations between the basic elements or sub-systems of a numerically controlled machining process as to the ways of the risk factor effect.

A numerically controlled machining process differs from conventional machining for example in that errors made on the different organizational levels may be conveyed through the process and finally cause a dangerous faulty machine movement. Therefore the study must also specify the working routines indirectly related to the machining process, and the critical points of the routines, as comprehensively as possible.

A numerical machining process is as a whole a very complex process. Therefore the research work starts from a systemic approach, in other words, the subject of study is divided into appropriate parts, which can be viewed fairly independently. In the course of the research these parts and their relative position become more and more specific, until a situation is reached when new material no longer changes the disposition and the critical points determined according to it. Thus the study in a sense proceeds through constantly emerging problems and their specifications, until new material in a particular sub-section is no longer obtained or new views can no longer be set forth. So the study does not start from a complete analysing method. It uses for example a failure mode and effect analysis, a safety analysis of working methods and other similar methods quite freely for analysing the diverse subsystems.

The study covered all machine tools equipped with numerical control. Of these, lathes were the most important group both in number and hazards involved. The most serious industrial accidents had happened with lathes, when the cutting tool or the holder had collided with the chuck or the chuck jaw. Due to the high rotation speeds of the chuck (even 6000 - 8000 rpm), a flying chuck jaw causes a great danger in spite of eventual guards.

Assessment of the critical components or the subsystems of the technical equipment is mainly based on assessments made by the users or the persons in charge of maintenance. There is also information available on the defects of some of the subsystems.

Errors in working methods have been specified starting from the view that incorrect operations are due to human errors made while regulating the working process. These adjustment errors are associated with defects in the psychical pattern of the working process or with the wrong use of that pattern (Hacker, 1978). However the actual wrong human action cannot be unambiguously derived from this theory. This would need analysing the reasons for the wrong action on the level of psychical regulation. Such results are not yet available within occupational psychology. Thus suggestions for minimizing faulty actions include measures that primarily concern improving the ergonomical properties of the equipment and the occupational skills.

HAZARDS OF NUMERICALLY CONTROLLED MACHINE TOOLS

The use of numerically controlled machine tools also involves industrial accidents of the same type that occur with conventional machine tools: cuts by chips, stumbling, dropping of workpieces etc. However, they seem to decrease in number with numerically controlled machining thus these accident types are left outside the scope of this study. This study concentrates only on such occupational accidents that are at least partly caused by numerical control.

Compared with the conventional types the numerically controlled machine tools are marked by certain characteristic risk factors. They are typically associated with entangling and cutting hazards caused by unexpected machine movements or hazards of flying objects in collisions. The potentiality of these hazards has grown for the following reasons:

- Numerically controlled machine tools are more complex than before. When the knowledge and skills of the operators cannot keep up with the new requirements, the potentiality of operational errors increases. Often not even the importer of the machine is sufficiently well acquainted with the construction and properties of it.

- The movements of numerically controlled machine tools may be fast and unexpected. The progress of machining cannot be directly seen, it can only be followed by display devices. Sometimes this does not give the worker enough time to react.

- The machines have become more powerful. The energy directed to the worker by faulty movements or flying objects can be very high, making the consequences of occupational accidents severe.

- The significance of organization is growing. Numerical control brings different groups of workers together in a common working plan. Errors can be carried hidden in the process so that in the end the operator may get into danger.

Studies of serious industrial accidents have proven that the greatest hazards with numerically controlled machine tools are

- collisions of moving machine parts with each other or with the workpiece, whereupon flying objects cause danger

- loosening of workpiece from its mounting and the resulting hazard of flying objects and

- other incorrect movements of machine parts and the entangling and cutting hazards caused by them.

On the basis of the studies on accidents and near accidents with numerically controlled machine tools, the indirect causes of accidents can be classified in the following way (Cf. also annex):

1. Errors in working methods and programming

2. Erros in operation
   - faulty setup
   - faulty operation

3. Errors in the machine system
   - functional errors in the control system
     - defects in the equipment
     - errors in the system software
   - defects in the operating system
   - defects in the mechanical system

4. Shortcomings in the safety equipment

Each occupational accident may involve several indirect causes. Every indirect cause is indicative of a critical point. Moreover, critical points can be recognized by the methods of risk analysis.

The potential effects of risk factors can be minimized by guards and safety equipment. As to large machine tools like vertical and shaft lathes, the machining space cannot be totally enclosed. Transfer lines, robot cells and FMS units cannot be completely barred either. Often the worker has to enter the operating area of the machine, in connection with programming, adjustments or trouble-shooting, while the machine is running or ready to run. Therefore the whole machine system should be designed and constructed so that the possibility of errors is at all stages as small as possible. In other words, one tries to eliminate critical points or at least minimize their possibilities of having effect. In the following chapters, some general problems and their solutions will be presented.

MINIMIZING ERRORS IN WORKING METHODS AND PROGRAMMING

Preparing a work includes choosing the way how the workpiece will be mounted, the cutting order and the tools as well as determining the cutting values. Errors made in the preparation of a work may remain uncorrected, which can result in a collision or loosening of the workpiece while machining.

Making up the work cycle, in other words programming the machine, is based on the preparatory work. Programming includes setting the tool paths in accordance with the cutting order, and giving other functional commands. There are three types of program errors:

Errors made in the preparatory work are manifested in the machining program as so called technological errors. Such are, for instance, incorrect turning speed of the chuck or incorrect cutting values. They can lead to too high a cutting force, so that the workpiece may come loose. Technological errors can be eliminated to some extent by automatic testing of the appropriateness of each functional command, for example the range of the programmed value. The consequences of these errors can also be reduced by surveying the cutting force or power.

The most common errors in a machining program are so called geometrical errors. Such are, for instance, wrong coordinates of the cutting tool position or incorrect cutting tool paths. Geometrical errors may lead to collisions. These erros can be eliminated by programming such prohibited areas in the machining space which the cutting tool cannot reach (the so called soft limits). In future the shape of the workpiece can also be taken into account so that during repositioning (quick transfer) the cutting tool passes the workpiece automatically (Fig. 2).

Syntax errors include, for instance, wrong characters or incorrect order of characters. These errors usually turn up automatically when the program is fed into the machines's memory. With automatic coding formal errors occur very seldom.

All errors mentioned above can be decreased considerably by moving from manual to automatic coding. This also makes programming faster and easier.

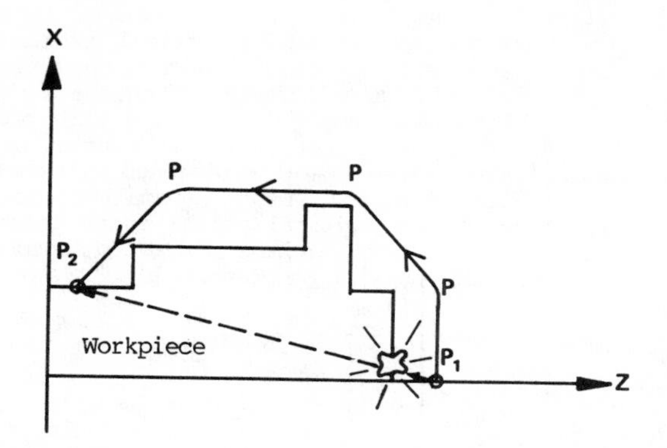

Fig. 2 Moving the cutting tool by quick transfer from point P_1 to point P_2. To avoid collision, auxiliary P points have to be used. In future it will be possible to design the motion line on the basis of the geometrical lines of the machine, cutting tool and workpiece.

MINIMIZING OPERATING ERRORS

There will, however, always be errors left in the programs. It is not uncommon that there are even several errors in the machining program for a complicated and new type of workpiece. Therefore the programs must be debugged before feeding. Possibilities of a collision may be detected in connection with programming by simulating the machine tool for example by a plotter of by checking the program by means of a debugging aid routine. A new program must always be checked before it is introduced, for example by running the program without a workpiece and at a safe distance from the machine parts or by machining the first workpiece at low feed and speed.

Programs are usually stored on punched tapes or magnetic tape cassettes. Defects in the filing system may cause different programs or different versions of a program to be mixed up, which again can result in a collision.

The operator of a numerically controlled machine tool gets the information on the progress of machining and the data on the machine mainly on a display device. The control system of the machine should give the vital data in an appropriate form and at the right moment. The control system should, if possible, anticipate malfunctions and display messages or alarm against them in time, and if necessary, stop the machine.

The dimensions and shape of the workpiece may vary within certain limits, but sometimes these tolerances are exceeded. In that case risk of collision may arise or the holding force securing the workpience may become too weak, which can result in a hazardous situation. A remedy for this is to check the principal dimensions of the workpiece before it is mounted on the machine. In future, an automatic work-changing device will be able to check the dimensions of the workpiece.

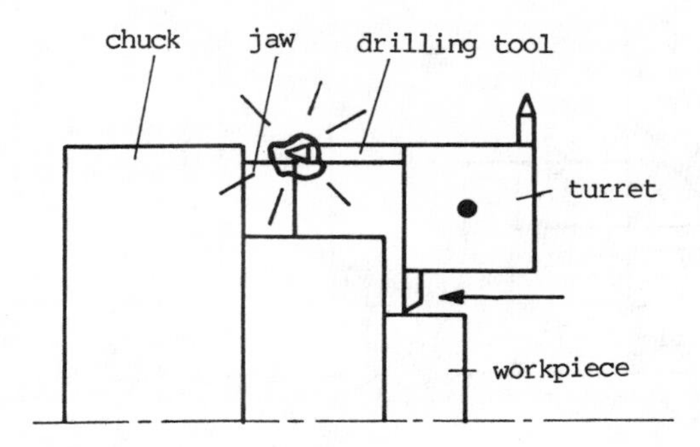

Fig. 3 Collision hazard of a long tool with the workpiece or the chuck in chuck work due to the incorrect tool setting length and position. The hazard can be avoided by changing the place of the long tool.

Errors in the preliminary tool setup may cause a collision, even if all the other work phases had been faultless. For example an idle long tool mounted on the turret may collide with the jaws of the rotating chuck (Fig. 3). Errors in setup can be eliminated by using a standard setup as often as possible. A tool in a wrong place in the tool magazine also causes a collision risk. This can be eliminated by a control system that identifies both the number of the tool and its place number.

SECURING AND INTERLOCKS

The reliability of numerically controlled machine tools was still inadequate in the 70's. Technical defects raised the average stoppage time to 5 - 10 per cent and the malfunctions in the machine often led to a collision. Today the reliability of computer-based control systems is already dependable. Electronic circuits are nevertheless vulnerable to external electrical disturbances. This can be managed only by careful shielding and securing.

The reliability of equipment can be increased by redundancy (doubling). However, this also makes the equipment more complicated and less practicable and therefore it is favourable to separate the safety functions from other operational functions. In this case, very high reliability is required only of the safeguarding (Fig 4).

The significance of the machine software (system software) is growing. By developing the software, the functions of the machines can be diversified and the operation can be made easier. The software can control the data from the machine as well as its own function. However, all this makes the software more complex and guaranteeing accuracy becomes a problem. Complicated programs cannot be fully tested and so software bugs are evident. Today, methods are being developed for working out programs that have a higher quality and thus program errors and their harmful effects will be decreased (Andersen, 1984).

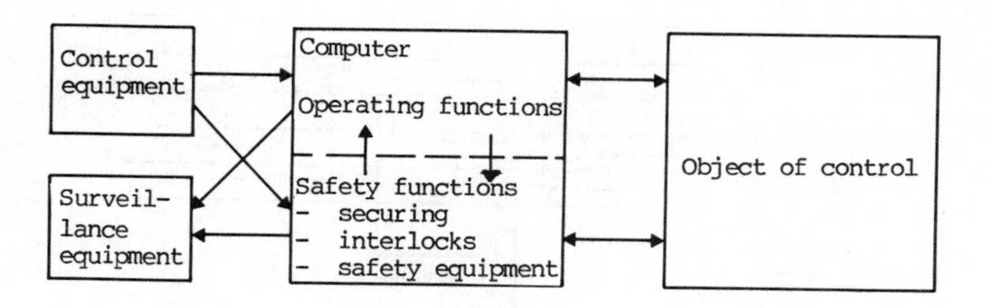

Fig. 4 Separating operating and safety functions from each other.

Increasing the reliability of the machine does not, however, solve all safety problems. By adding extra controls that are independent of the operator, the machine's functions can be secured and risk factors prevented from developing. An example of this is securing the mounting of the workpiece during machining (Fig. 5).

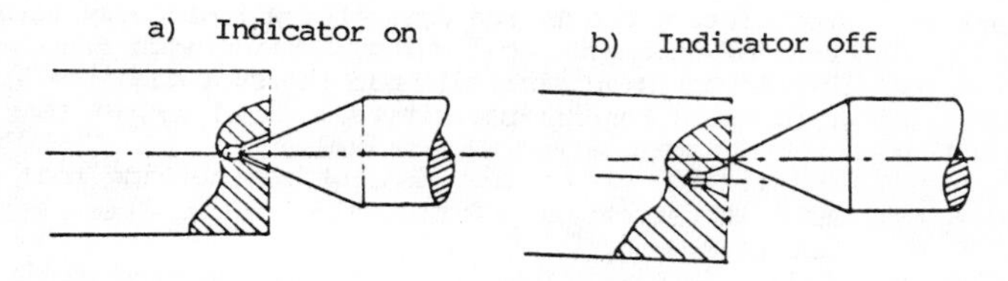

Fig. 5 Adjusting the tailstock centre to secure the mounting of the workpiece. In the case in figure b the machine will not start (SMT).

Errors in operation and especially their effects can be minimized by interlocks. Risk of collision can be lessened by restricting the operating areas of the moving machine parts (Cf. Fig. 6).

The surveillance of the machine contributes to preventing major maintenance. This also adds to the safety, because a great deal of the defects can be discovered and eliminated before the actual fault comes up. For instance, the most important parametres of the machine can be continuosly followed by sensors.

a)

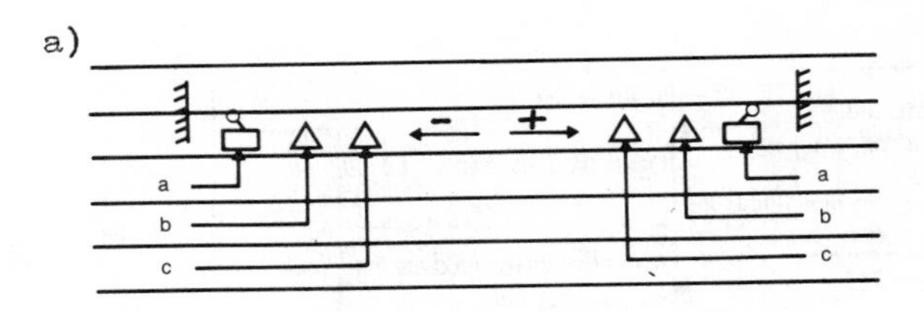

b)

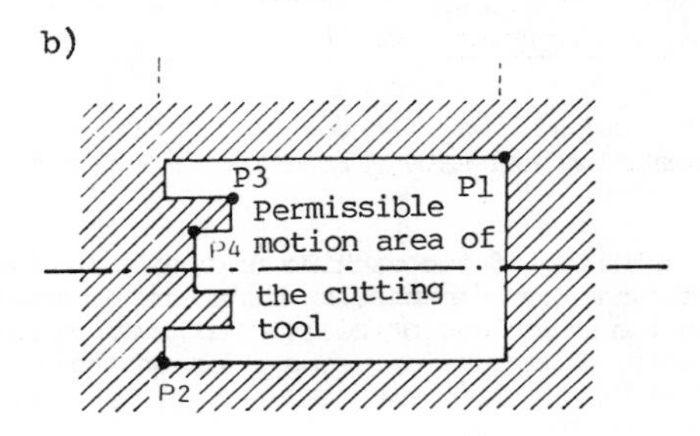

Fig. 6 a) Restricting the motion area of the tool-carrying slide.
a-a: electromechanical safety limit switches.
b-b: programmed limits in machine system.
c-c: user programmable limits.

b) Restricting the motion area of the cutting tool by program mable safety areas (Okuma). The area has been determined by points P1 - P4.

GUARDS AND SAFETY DEVICES

A prerequisite for securing the safety of automatic machines is that the machine will function in the intended manner. To prevent or minimize the effects of the potential hazards of the machine adequate guards and safety devices must be provided.

The most economical solution is to isolate the risk factor by a fixed or movable guard covering the dangerous area as closely as possible (enclosures, telescopic guards etc.) If this is not possible, safety devides connected with the functions of the machine shall be used (see Figure 7). They include:

- safety gates or hatches etc.
- automatic reversible operating devices
- double-handed operating equipment
- photo cells
- safety mats and linings, etc.

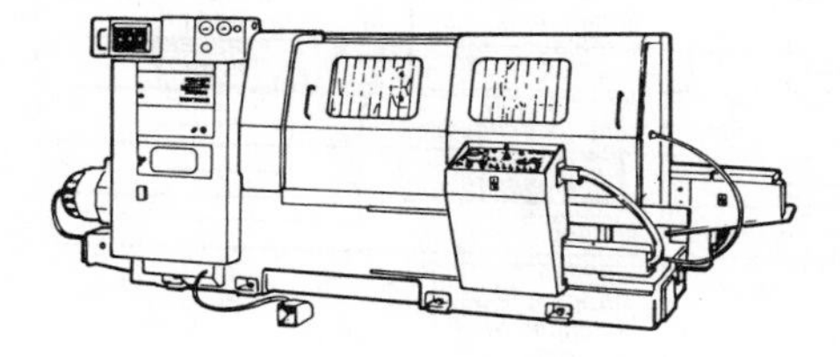

Fig. 7 A numerically controlled lathe with a completely enclosed machining space (Takisawa).

The connection of a safety device with the operating system must be reliable. If the operating system is doubled the connection of the safety device must also be doubled. Simple control systems must be secured by the use of reliable components. Limit switches must be of forced action type (safety limit switches) and operation relays must be cross-connected so that sticking points will be detected.

TOWARDS SAFE MACHINING

The introduction of numerical control in machine shops does not as such improve occupational safety. The control systems often lack vital securings and machine control functions. The safety-technical solutions are still inadequate and when automation is applied, the worker is often left with only monotonous subsidiary duties.

Nevertheless, numerical control offers quite new opportunities for improving occupational safety and it also develops the content of work. By constructing safe, reliable and convenient machine tools as well as combining them into flexible production systems, it is possible to discard dangerous, heavy or monotonous work routines. It is also possible to unite programming, setup and repair and maintenance work into such meaningful entities of work, which to not hamper the operator's development at work. This, however, requires versatile occupational skills. The present system of vocational training has to be developed so that we can make the most of the opportunities new technology offers.

REFERENCES

1. Sundquist, M.: Numeerisesti ohjattujen työstökoneiden työturvallisuus (Safety with numerically controlled machine tools). Tutkimusraportti no. 45. Työsuojeluhallitus, Tampere, 1983 (in finnish).

2. Hacker, W.: Allgemeine Arbeits- und Ingenieurpsychologie. 2. Aufl. VEB Deutscher Verlag der Wissenschaften, Berlin, 1978.

3. Andersen, O.: Personsikkerhed ved mikroprocessorstyringer (en delraport). Elektronik Centralen, 1984. Denmark.

Annex

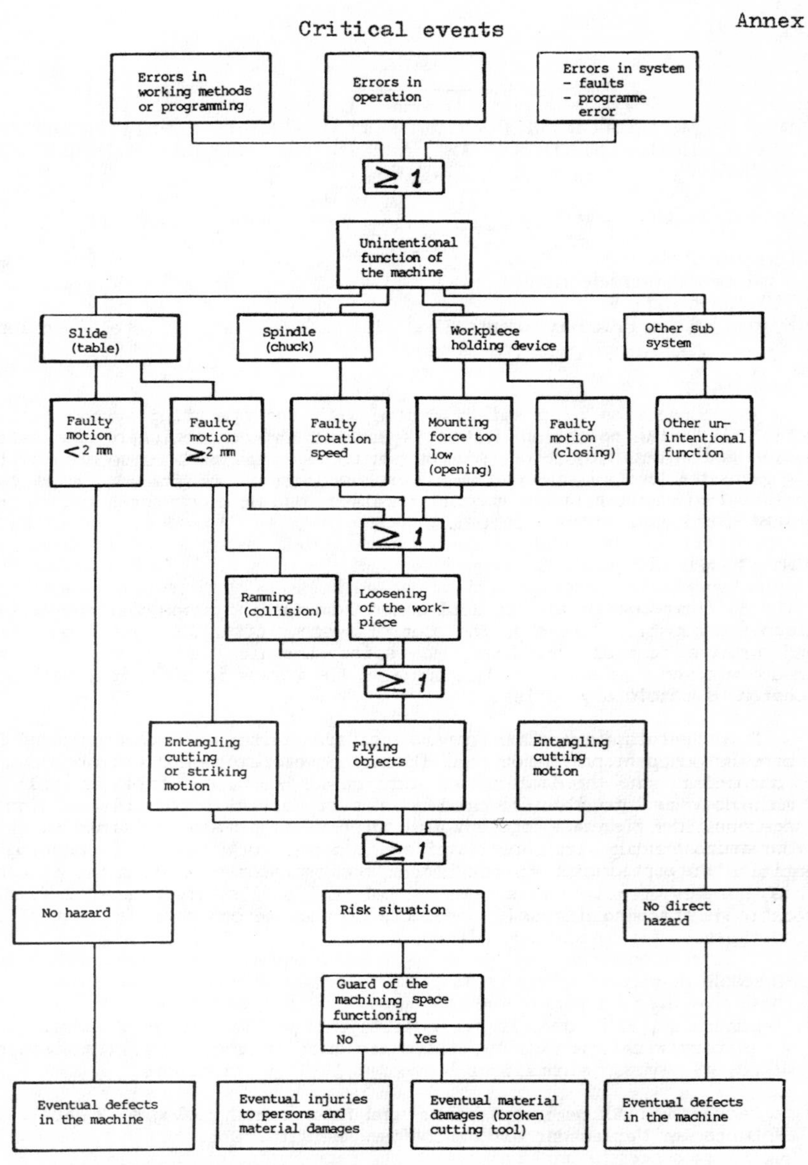

Model of mechanical hazards on a numerically controlled machine tool. The model describes the typical flow of incidents, which can result in injuries to persons and in material damages.

REQUIREMENTS FOR MICROCOMPUTER SYSTEMS IN SAFETY RELEVANT APPLICATION - STATE OF THE ART IN THE FEDERAL REPUBLIC OF GERMANY

Dr.-Ing. Jansen, Herbert

Director of the
Institute for Microelectronics
TÜV Rheinland e.V.
D 5000 Köln 1, P.O. Box 10 17 50

ABSTRACT

Technical safety in Germany is postulated by legal provisions and explained in detail by recognized rules of technology. For control or protective systems the level of safety mostly is explained by fault flow charts. The direct application of these failure theories for programmable electronic systems is in most cases not possible. It was therefore necessary to collect the existing experiences in the field of computerized safety systems and to create a structure, which meets the requirements of the rules of technology and on the other hand makes practical application and licensing of these systems possible. The TÜV handbook is the result of such studies and is the first step into this new territory.

INTRODUCTION

Technical safety in Germany is legally expressed by laws, ordinances to laws and accident prevention regulations. The actual technical situation can be classified by the three categories state of science and technology, state of technology and generally recognized rules of technology the first of which describing the highest - possibly still theoretical - level, the second stating the technologically realisable level and the third explaining a minimum but generally accepted level according to technical standards.

These technical standards are DIN, VDE, VDI or other standards.

Safety objectives or goals of legal provisions are work protection, household or private sphere resp., traffic safety and environmental protection. These provisions are part of the administrational and social law.

Safety itself is defined in the technical standard DIN V 31004 as a situation with less risk than a given risk limit.

Safety cannot be measured in a straightforward approach, but can only be explained by the complementary expression "risk". One does not count the number of successful missions, but rather lists up the accidents! So safety may be explained as a "situation with sufficiently low risk level". Absolute safety cannot be achieved.

The subject on which we shall focuse our attention during this session is a "technical safety device" related to a technical system or process. These devices are generally based on conventional technology, which means relay circuits or discrete/integrated electronics. But there are lots of applications in Germany already implementing microcomputer-controlled safety devices.

1. Requirements for technical safety devices

Risks result from technical processes (nuclear power plant, railway system, steam boiler, elevator, invasive medical apparatus). The risks of the technical processes may derive from a malfunction of the technical process itself or from a malfunction of the safety device. This safety device either works as a protective system, which means that it monitors the technical process and danger may arise out of the process, if the monitoring function is lost through a failure. The safety device may work as well as a control system, which means that it actually influences the technical process and might, when working in the wrong way, give wrong or dangerous commands to the technical process. To rule out that safety devices do not work correctly, they must be fail-safe. This aspect will be explained later.

1.1 Fundamentals of fail-safe design

Regarding the functional technical safety, the safety device has to be designed to cover the requirements of the technical process or system concerning the fail-safe conception and relevant safety level.

The first step is to analyze the technical process or system in terms of safety and define the so-called safe situation. The safe situation is a situation which, by definition, is equally or even more safe than the actual situation and can be achieved immediately out of the actual situation by taking simple action.

E.g. the emergency stop in traffic systems is a simple action, leading the system (train, cabin) to a safe situation. Actually, there is no real safe situation in the above-stated sense for most technical systems: imagine an underground train catching fire. In this case it should not stop, but try to reach the nearest station. The automatic safety device should not be fail-safe, but keep the train running. In other cases it makes sense to stop the train. So the safety device should be intelligent to make the decision. But how to design intelligent, non-fail-safe but well fail-safe devices if necessary? So this problem leads to the so-called "by definition" safe situation.

It is easily understandable that the level of safety and thus the costs of a safety device should be adapted to the characteristics of the technical process. If e.g. the process is slow enough, the safety device may issue a faulty command if this error is detected in time. Quick processes call for error-free commands.

The second step is a detailed failure mode and effects analysis (FMEA) for those sub-systems which have to be designed in a fail-safe way according to the findings obtained from the systems analysis.

In Germany fault flow charts are normally used for this second step.

1.2 Fault flow charts

There are various fault flow charts in standards or other rules of technology, covering the specific design requirements of fail-safe behaviour for the safety device in question. The principle of all fault flow charts is always the same. The chart begins with the "1st failure" (e.g. emitter-collector of any transistor short circuit). It is to be verified that after each "1st failure" no dangerous situation may occur. If so, one has to ask what else happens after the "1st failure". There are 4 answers to this question:

- The "1st failure" immediately leads to the "safe situation". If any first failure runs this course, the safety device is fail-safe to the highest possible level. Only relay switching circuits or small discrete electronic systems can be designed this way. This conception also adversely influences the reliability.

- The "1st failure" leads to signalling (operator informed). In this case it is up to the operator to decide what to do.

- The "1st failure" will not be noticed immediately, but will be detected during the next system inspection (automatic check or manual test). In general it is not assumed that during the interval between failure appearance and failure detection any "2nd failure" will occur and thus, in combination with the "1st failure" create a dangerous situation. It is often a question of probability calculation depending on the length of the interval, if this assumption is permitted.

- The "1st failure" will not be detected at all (not detected by routine inspections).

  In the last mentioned case, the "1st failure" has to be combined with any "2nd failure" and the flow chart is followed for the same routine of dangerous (not allowed) or safe situation or signalling or detection by inspection or non-detection.

  According to the level required, this may be repeated infinitely, e.g. in railway signalling systems (VDE 0831/DIN 57831).

2. Realization of computerized fail-safe circuits (theoretical background)

From the use of computers for safety relevant applications new problems arose due to the different technological and structural design of computers compared with hardware logic.

The electronics of computers is only able to operate with programs - software. Besides the aspects of hardware, which have to be regarded in conventional control devices as well, also specific software-related issues have to be solved.

Even simple computers have a very complex design.

The technological reason for this complex design is the integrated technology and the large number of internal memories. Compared with conventional control devices, also the software is very complex:

The large number of input combinations and the huge amount of different software paths create situations which are hard to follow. Failure exclusions for computers are not reasonable. Failure effects are dependent on program, data flow and time and thus undefined. A fail-safe behaviour of single computers cannot be achieved.

2.1 Safety of computer devices

2.1.1 Pre-operational test

A complete test of a computer has to cover all commands with all address and data combinations and all possible successions of these combinations. 200 different commands and 2^{16} addresses and 2^{16} data combinations lead to about 10^{12} test patterns. Assuming 100 tests per second (computer-aided), testing will take 320 years.

If the computer including its software has to be tested as black box, then 2^{m+n} test patterns for testing of the software are necessary, where m is the number of binary input variables and n the number of binary internal memories.

For m=16 input variables and n=64 internal memories we find 2^{80} different combinations. With 100 tests per second, testing will take 4×10^{14} years.

So we find out that a complete test is impossible to conduct.

2.1.2 Failure modes and effects

Computers have failure rates of 10^{-3}/h up to 10^{-5}/h. Failure exclusions are not allowed with failure rates like these. Effects of failures cannot be predicted as they

- depend on the programs
- depend on the data
- depend on time of occurrence.

Due to the number of elements, a FMEA as explained above cannot be undertaken on the basis of internal computer analysis.

Failures may not be recognized immediately. It is even possible that failures in rarely used parts of memories or failures depending on data flow remain undetected for a long time and may lead to an accumulation of failures.

2.2 Conclusions for the safety of computer devices

2.2.1 Pre-operational situation

Hardware

If there is no possiblity of testing computers, one should use several hardware channels.

2 channels are used for systems with a safe situation, 3 channels for systems with no safe situation.

The channels are non-uniform, different computers from different manufacturers with different technology are used. From this it can be expected that there are no identical errors in the different channels.

In single channels fail-safe voters detect failures and initiate the safe situation. Otherwise the defective channel is excluded by a majority voter.

The level of diversity desired or possible depends on the safety level required by the technical process.

1 channel solutions for lower safety levels are also possible with restrictions explained below.

Software

As for software, the same diversity as explained for hardware is required. The connection of the diverse software channels has to be achieved by small, fully tested programs or hardware.

If no software diversity is required, the "1 channel software" has to be correct. Correctness is proved through analysis and test.

2.2.2 Failure modes and effects

Failures cannot be excluded. So they must not lead to a dangerous situation and have to be detected. This can be done by comparison

- with a desired value, e.g. stored in a memory for computer self-check
- between different channels.

Again we make a distinction between protective an control systems.

Protective systems

With a 2 channel protective system a failure in one channel will not prevent the protective measure from taking effect. The failure is detected. A computer self-check assures data flow-independent failure recognition.

This system is fit for non-fault tolerable processes.

1 channel protective systems with self-check and extraneous tests are also possible. The self-check has to find out all failures, if possible. The extraneous test provides for a testing of the basic functions of the self-test. As the self-check takes some time, this system is only applicable for temporary fault tolerable technical systems.

Control systems

For a 2 channel system for processes with safe situation, the control signal is derived from channel 1. Channel 2 operates as comparison to initiate the alarm and failure signal. The alarm leads to the safe situation of the process. Also very quick technical processes can be controlled by this system.

A 1 channel control system for processes with safe situation again needs a self-check. This self-check has to detect all failures if possible. The extraneous test provides for a testing of the basic functions of the self-test. This system is only applicable for technical processes with a fault tolerability which exceeds the failure detection time.

Control systems for processes without safe situation

3 channel system: The system has 3 channels which lead to the process via a 2 out of 3 voter. A failure in any one channel does not effect the control signal, but generates a failure signal. Data flow-independent failures have to be detected by self-checks within the computers. The system is fit for quick processes.

2 channel system: The system has 2 channels and works as hot stand-by. Each channel is fitted out with a self-check to find out all failures, if possible, whereas the extraneous test provides for a testing of the basic functions of the self-test.

If a failure in the channel on duty is detected, the stand-by channel is connected with the process. The switching must be fail-safe.

Again this system is only applicable for temporary fault tolerable processes.

3. The TÜV handbook (practical approach)

The said requirements for microcomputerized safety devices were worked out in Germany mostly by TÜV Rheinland for a broad area of application, using all the information gathered by cooperation with the Federal Railroad Organization, the professional associations for work safety, the commission for nuclear safety and many other relevant groups or authorities. Practical experience in using this knowledge was gathered by type approval and licensing of computerized safety devices for various purposes.

The highly sophisticated system of standards in Germany with its fault flow charts as explained above cannot easily be adopted for computerized systems. So there was a need for technical help for design engineers and manufacturers - which was achieved by the TÜV handbook for microcomputers.

The work which forms the basis for this book was carried out under the research and development project "Microcomputer in Technical Safety", supported by the Federal Minister for Research and Technology, by the TÜV study group on computer safety (TÜV Rheinland/TÜV Bavaria).

The following is a literal quotation from the introduction and guide to the use of this handbook:

The legislative provisions which are currently in force do not cover - or do not cover adequately - the special circumstances/conditions which are relevant in the employment of microcomputers. It is often a matter of some difficulty to fulfil the requirements laid down in various rules and regulations as to the behaviour of equipment in the event of a fault without incurring expenditures which cannot really be tolerated; in some cases therefore the requirements are not met to an adequate extent. Where microcomputers are employed in fields which are related to safety therefore, it is not simply a question of moving into new territory as far as technology is concerned, but also as far as safety certification is concerned.

The task of this handbook is to make the journey into this new territory easier for those who design/develop or manufacture safety systems and equipment. It should help them to design and build their products in such a way that they conform with all the generally recognized technical rulings. This handbook however does not contain any simple "formulas" nor does it contain any novel provisions which set out to replace or to extend those which are presently in force. It contains, rather, a catalogue of system layouts/structures and safety measures for microcomputer control systems, and shows the way to select those safety measures which will ensure that, in a given situation, the requirements of the relevant legislative provisions will be complied with. There are in this connection a number of possible options from which the designer or manufacturer can choose, in accordance with the operating conditions which he has specified for his product, so as to provide the optimum solutions for the problems in hand.

Unquote.

The book is based on the fault theories of the technical standards and classifies into 5 categories all safety relevant applications according to the number of faults, in the presence of which no dangerous situation is accepted. Examples and relevant standards are given for each category. On the other hand computer structures and measures to be employed in order to meet the safety level of the individual category are listed up.

Various structures and measures can be combined for equivalent individual measures. As an example, category 3 for two possible 2 channel and 1 channel solutions is listed below:

Category	Structures and measures against Failures	Structures and measures against Systematic faults	Tests
3	2 channel, comparison of results complementary tests	operationally proved and certified components	checking of required measures and characteristics, coding inpection, checking of all program functions, checking of all time constraints and breaks, complementary testing of functions on completed equipment
3	ditto	ditto	checking of required measures and characteristics, manual program analysis, checking of all time constraints and breaks, complementary testing of functions on completed equipment
3	1 channel, high level tests or monitoring measures for - ROM - RAM - Input/Output, - CPU. Time-based and logic-based monitoring of running of program	operationally proved and certified components	intensive checking of required measures and characteristics, theoretically, practically - by detailed fault simulation, etc., coding inspection, checking of all program functions, checking of all time constraints and breaks, complementary tests of functions on completed equipment
3	ditto	ditto	intensive checking of required measures and characteristics: theoretically, practically - by detailed fault simulation, etc., manual program analysis, checking of all time constraints and breaks, complementary tests of functions on completed equipment

The book is not a recognized rule of technology. But as a first result, the DKE Standard Committee K 714 "Programmable electric systems for safety relevant applications" was founded in order to issue a DIN/VDE-Standard covering these questions.

CONCLUSIONS

The TÜV handbook was the first step for a general structure of measures to be taken for the safety of computerized electronic devices covering all fields of technical application. Its broad acceptance in industry and test centres has shown that there is an urgent need for concepts of computers in control or protective functions.

In Germany many organizations and companies try to improve the safety of computerized systems - one of the inputs for this work is the TÜV handbook.

REFERENCES

1. Hölscher, H., Rader, J., Mikrocomputer in der Sicherheitstechnik, Verlag TÜV Rheinland, Köln, 1984

2. Safety Related Computers, European Workshop on Industrial Computer Systems, Verlag TÜV Rheinland, Köln, 1985

3. Meffert, K., Germer, J., Einsatz von Rechnern für Sicherheitsaufgaben - Standortbestimmung. Die BG 5, 1985, 246.

USE OF MICROPROCESSORS IN SAFETY CRITICAL APPLICATIONS - GUIDELINES FOR THE NORDIC FACTORY INSPECTORATES

JØRGEN BØEGH
OLE ANDERSEN
SØREN P. PETERSEN

ELEKTRONIKCENTRALEN
Hørsholm, Denmark

ABSTRACT

A set of guidelines developed for the Factory Inspectorates of the Nordic countries are presented. The guidelines apply to development of microprocessor based systems used in safety critical applications, and their purpose is two-fold. First of all, the guidelines shall give constructive advice to designers and manufacturers. At the same time, they are to be used by the Factory Inspectorates as a basis for safety analysis of the systems.

INTRODUCTION

ELEKTRONIKCENTRALEN has prepared a report [1] for the Nordic Council of Ministers (Nordisk Ministerråd). This report will serve as background material for the development of a set of guidelines aiming at enhancing the level of personnel safety in microprocessor applications. The assessment methods and techniques applied to traditional systems are primarily based on experience accumulated over the years. Unfortunately, the error pattern for microprocessor based systems are different from those of traditional systems. Since the development in microprocessor technology is so fast, it will be impossible to rely exclusively on experience in the safety assessments. Hence, it is necessary to base assessment guidelines on more general principles. The framework proposed and the ideas behind this framework are described in the following. Much of the work is inspired by the results of the CEC collaborative project, in which ELEKTRONIKCENTRALEN has participated.

THE AIM OF THE GUIDELINES

The aim of the guidelines for safety assessment of microprocessor based systems is two-fold. They must fulfil the needs of both the manufacturer and the assessor. The guidelines will be of maximum use for the manufacturer if they are operational, i.e. if they contain constructice advice for the designer and for the producer. It is important that the designer knows the requirements of the assessor at an early stage in the design process. Thereby, the designer gets the possibility to integrate the safety aspects into the system in a natural way. This will result in better (more safe) products and also ease the assessment procedure. At the same time, the guidelines will be used by the Factory Inspectorates in the Nordic countries as a basis for safety analysis and assessment of the systems.

It should be stressed that even if both the manufacturer and the assessor strictly follow the guidelines, there will be no guarantee that the system in question is 100% free of errors. At present, there is no method that allows this level of confidence to be obtained for software in typical industrial applications, which is due to the rather limited resources available for system development and verification. Instead, the guidelines aim at forcing both the manufacturer and the assessor to perform a disciplined system analysis, by braking down the task into a series of specific actions. General experience shows that this can be expected to lead to systems of high safety integrity. A reasonable goal is that of microprocessor based system reaching safety levels which would be acceptable for similar traditional systems.

THE GENERAL FRAMEWORK

When a general framework for safety assessment is established, a number of points must be kept in mind. Of particular importance are the following requirements:

- the assessment must be performed in a systematic way
- some flexibility in the choice of analysis method is allowed
- future technological development can be utilized
- there must be good agreement with traditions for both system development and safety assessment
- the greatest possible international agreement on the assessment methods must be obtained

The general framework for safety assessment proposed takes these considerations into account and follows closely the framework developed within the CEC collaborative project. The framework splits the assessment procedure into seven well-defined steps, which should be worked through in the given order. However, it might turn out during the assessment procedure that a previous step has to be reconsidered. Then care should be taken to incorporate this additional information into the analysis performed in the other steps.

The seven steps of the general framework have the following contents:

Step 1: System Boundary Definition
It is examined of which hardware and software components the system consists. Furthermore, these areas are identified where the system can cause danger, both when in normal modes of operation and in case of a malfunction.

Step 2: Hazard Analysis
The purpose is to identify the hazards of the system and to determine events leading to hazards. For each identified hazard, its consequences and probability of occurring are determined.

Step 3: Safety Requirement Specification
The system is classified according to its application area. On the basis of the results of the hazard analysis (Step 2) and specific legal requirements, the safety requirements to the system are specified. For the manufacturer, it is important that the requirements are established as early as possible in the design process, in order to avoid costly rework later in the development.

Step 4: Identification of Safety Critical Subsystems
The boundaries obtained in Step 1 are restricted to the safety critical subsystems. This step concentrates the further work to the safety related part of the system.

Step 5: System Analysis of Critical Subsystems
In this step, a detailed analysis of the safety critical subsystems is performed. The purpose is to verify that the system complies with good engineering practice and that it implements the safety requirements specified in Step 3. This step accounts for most of the work in the assessment procedure.

Step 6: Certification Criteria
The results of Step 3 (safety requirements) and Step 5 (system analysis) are compared, and it is judged whether the system is sufficiently safe. In this step, common sense and general knowledge of technology and safety problems must be utilized in determining whether the system can be certified.

Step 7: System Change Management
When a system is certified, procedures for handling system changes must be established to ensure that no degradation of safety occurs during the life time of the system.

SAFETY ANALYSIS

In Step 3 of the general framework it is determined which safety properties the system has to satisfy and how to verify these properties. The detailed verification takes place in Step 5. The guidelines distinguish between five different assessment methods:

- Study of Documentation:

  An experienced person works through the documentation in order to demonstrate the desired properties

- Inspection of the System:

  The hardware and software of the system is considered, e.g. code walk through.

- Answering Checklists:

  Checklists exist for hardware, software, and architecture aspects of the system. Answering a question may demand the use of any of the assessment methods. The answer can be of the type: YES, NO, NOT APPLICABLE, or a scoring in a specified interval.

- o Analysis and Calculations:

  A number of methods are available, e.g. fault tree analysis, failure mode and effect analysis, stress analysis, worst case analysis.

- o Test:

  The purpose of testing is to demonstrate properties which are difficult or impossible to verify theoretically, e.g. EMC properties and complex software functions. It should be noted that software testing in itself can never prove the correctness of the software.

Some of these analysis methods are also applied in Step 2 (hazard analysis). In the analysis, seven safety quality attributes are used. They are divided into two groups.

- o Primary Quality Attributes

  The primary quality attributes are properties aiming at the avoidance of errors:

  - High Reliability
  - Safeguards against handling errors
  - Safeguards against intended misuse

- o Secondary Quality Attributes

  The secondary quality attributes are properties which should prevent an error from resulting in a safety critical situation:

  - Fault detection
  - Fault correction
  - Fail to safe
  - Fail operational

As far as possible, the questions on the checklists are split into these seven categories. The questions covering each quality attribute reflect the possible error sources. The error sources can be structured as follows:

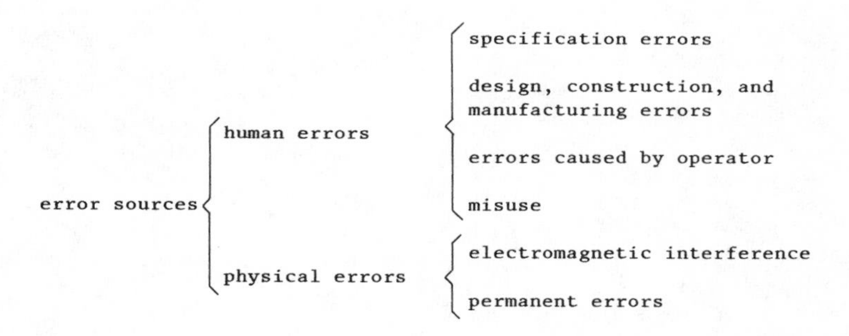

The questions on the checklist are at the same time a source of good advice to the producer of the system, since the questions attempt to determine whether good engineering practice was followed during development. In addition, the guidelines contain a number of methods and techniques which are recommendable for the designer to help him avoiding these error sources.

The safety assessment of microprocessor based systems requires detailed knowledge of a broad spectrum of technologies. Since one person can hardly cover the entire assessment procedure, the main safety analysis is divided into three parts:

- o Software
- o Hardware and EMC
- o Architecture

Both the checklists and the design recommendations follow this technology determined division.

CONCLUSION

The essence of the framework and guidelines for safety assessment lies in its way of structuring the assessment procedure and constructiveness of the checklists. In this manner, the guidelines will be of maximum use in helping both the assessor and the manufacturer to obtain the highest possible level of personnel safety.

REFERENCES

[1] Andersen, O., Bøegh, J., Petersen, S.P., "Personsikkerhed ved mikroprocessorstyringer". ECR-report 1986 (to be published).

STANDARDISATION FOR COMPUTER SAFETY - THE CURRENT SITUATION IN GERMANY

K. MEFFERT

Berufsgenossenschaftliches Institut fur Arbeitssicherheit,
St. Augustin, Federal Republic of Germany

INTRODUCTION

A multitude of standards and regulations in Germany cover the safety of equipment installations in which computers are also to be found. Up to now, however, computer-specific problems, see for example (1,2,3), have been given insufficient attention. Protection against accident contact, i.e., protection against electric shock, operational safety, e.g., how machine locking mechanisms must be designed, and very generally, the behaviour of the installation should a fault arise are dealt with by this regulating framework. Whilst it is not possible to differentiate between a conventionally-designed control system and a computer system with regards to protection against accidental contact and operational safety (thus illustrating that computer technology does not enjoy special status in this case) this is not quite so when it comes to safety in the event of a fault in the system. Application-specific safety regulations often require that when a fault occurs certain potentially dangerous situations must not be allowed to arise and that faults must be automatically detectable in a certain way after they have occurred. In the field of conventional electro-mechanical control circuits and discrete electronic systems the experts were largely agreed on which faults are to be considered as likely in practise and how these faults can be detected through appropriate circuitry. Furthermore, fault elimination was found to be possible in a variety of switching elements as long as these exhibited certain permanent structural features.

The situation proves to be completely different when it comes to computer installations. Experts are by no means agreed on which faults are to be considered likely and how faults can be avoided or remedied. The software - a completely new element in computer-controlled equipment - also raised new problems in the safety issue whereby only the checkability of the system and the ease of modification are to be mentioned here.

This dilemma lead the Technical Control Associations (TUV) of Bavaria and Rhineland to form a TUV study group "The use of Microcomputers in Safety Technology". As part of a research and development project carrying the same name a project was financed by the Federal Ministry for Research and sponsored by the VDI technological centre in Berlin. The objective of the project was to provide guidelines for developers and manufacturers wishing to use microprocessors in the area of safety technology. The results of the project are summarised in a manual (4).

A manual of this kind can well provide guidelines for the developer but it cannot replace a general technical regulation or standard. This was why the German Electro-Technical Commission (DKE) formed a new committee K174 with the title "Basic Principles for Computers in Safety Systems". Thus committee has the task of preparing concrete measures to ensure the correct functioning of computer systems designed for operational and, in particular, safety applications. This includes the systematic description of procedures for the creation, checking and documentation of software, firmware and hardware including the transmission paths.

THE WORK OF DKE-K174: BASIC PRINCIPLES FOR COMPUTERS IN SAFETY SYSTEMS

This section is aimed at showing the state of activities undertaken to date. Initially, a short survey was carried out on the level of technology attained so far in the use of computers in safety circuits. Concrete implementations could be noted in the following areas:

- Fuel engineering
- Traffic engineering
- Reactor protection systems
- Machine tools
- Stage control
- Medical technology
- Others

Eight different applications were noted in the area of fuel engineering. The control structure ranges from single-channel through partly redundant to completely dual-channel solutions.

In the traffic engineering sector various control tasks were noted with reference to railway signal engineering (signal cabins, train control), tracked busses and cableway control systems. The solutions shown here are all dual-channelled or have two-out-of-three circuits; diverse redundant solutions have also been selected in some cases. A total of three examples for stage control systems were noted all of which had at least partial redundancy. Technical implementations in the area of machine tools were single-channelled with partial redundancy; two other implementations had complete diverse redundancy.

Systems which up to now has been running as an experimental reactor with two-out-of-three selection circuits with non-diverse redundancy. Software diversity is also planned for 1986. In the area of medical technology applications in dialysis machines, infusion pumps, infusion regulators were noted. There were a total of nine new implementations. The basic solution possibilities range from single-channelled technology with partial redundancy through to dual-channelled technology. Further implementations in refuse compactors, printer control systems and stored-program control systems with general safety applications could be noted ranging from single-channel through redundant to two-out-of-three selection circuits but in this case without diversity.

Attention is now drawn to the actual study concept of the K714 committee. The approach follows relatively closely the known concepts applied in conventional and discrete electronic engineering. Initially, it must be stressed once again that the behavioural aspects of installations when faults occur form the basis of all considerations and calls for new regulations (Fig. 1). In this context there are faults which arise before commissioning but only become apparent during later operation such as occur during planning, design or realisation. Software errors or systematically integrated weak spots in the hardware are typical here. Other faults arise later during operation such as random component failure or interference. Faults caused through alterations or faulty repair also belong to this category.

<u>Faults</u>

Arising <u>before</u> commissioning (usually systematic)	Arising <u>during</u> operation (usually random)
Faults arise in the following phases:	Examples of why faults occur:
- Planning/specification	- Failure
- Design/development	- intermittent failure - static failure (- systematic failure) (- single, multiple failure) - common mode failure (- sudden failure and drift failure)
- Realisation	- Interference
	- False input information
	- Alteration, faulty repair

Figure 1

The underlying idea can be formulated as follows: certain measures such as quality control, checks, organisation, facilitate the avoidance of faults before commissioning i.e., faults simply do not arise or are at least discovered and remedied before the equipment is handed over to the customer. The more superior and effective these measures are, the fewer faults remain in the product when supplied. In the ideal case these measures would be selected in such a way before commissioning that all faults could be discovered and remedied enabling the equipment to be handed over to the customer completely fault-free. Absolute freedom from error cannot, however, be guaranteed in systems using computers.

Those faults which cannot be avoided or occur at a later date during operation must be remedied. This means that these faults may not be dangerous and, as far as possible, that the faults are automatically detectable. There are a series of technical and non-technical measures which are suitable for remedying these unavoidable faults during operation such as redundancies, fail safe devices and regular repetitive checks.

Small study groups are currently occupied with compiling these

- measures for avoidance of faults and
- measures for the control of faults

and with assessing their effectiveness.

CONCLUSIONS

In conclusion attention must be drawn to another important aspect. The measures for avoidance and control of faults must of course be adapted to, and be suitable for, the case of application. It would be ill-advised to implement just one, albeit superior, combination of measures for every safety application - to say nothing of the practical realisation. It will therefore be one of the most important and difficult tasks to put together combinations of measures which are appropriate to the risk of a considered application. It will thus be necessary to introduce safety classes and to allocate appropriate technical and non-technical measure combinations to every safety class.

REFERENCES

1. DIN VDE 0160 Ausrustung von Starkstromanlagen mit elektronischen Betriebsmitteln.

2. DIN VDE 0800 Fermneldetechnik.

3. DIN VDE 0804 Fernmeldetechnik.

4. Holscher H., Rader I. Mikrocomputer in der Sicherheitstechnik, Verlag TUV-Rheinland, 1984.

HARMONISATION OF SAFETY STANDARDS FOR PES

B. K. DANIELS

European Workshop on Industrial Computer Systems,
Committee on Safety Security and Reliability,
and
The Safety and Reliability Society, PO Box 25,
Cambridge Arcade, Southport, PR8 1AS, UK

INTRODUCTION

Harmonisation is a catchword which has many shades of meaning. In this paper the development of the European Workshop on Industrial Computer Systems, its Committee on Safety Security and Reliability (EWICS TC7), and its links with standards creation is described. The work of this group is towards pre-standards adopted by industry as the users of systems, and to assist in the preparation of international standards in this difficult field.

Harmony is achieved at a personal level by the international collaboration of experts attending EWICS TC7 meetings. At and between industrial users of safety related computer systems, harmony arises through the EWICS members participating in the creation of guidelines and then applying them in their workplace.

In accepting input from EWICS TC7, standards bodies have access to well developed and supported guidelines that have generic application to many industries and uses of programmable electronic systems (PES). Another shade of meaning for harmony.

This symposium has adopted the term PES, and whilst the term has not been used by EWICS TC7 in its published work, it is synonimous with Industrial Computer System. The paper relates and harmonises the work of EWICS to other papers in this symposium, and to the extensive links established by EWICS with other bodies active in this field.

The work programme of EWICS TC7 is described, and its relationship with the Safety and Reliability Society and the Commission of the European Communities.

FORMATION OF EWICS

In the 1970's a series of workshops were held at Purdue University, West Lafayette, Indiana in the USA. The initiator was Professor Ted Williams. He identified the need for consultation between manufacturer and user of computer systems across a wide range of industries, and conceived the workshop series as the means of satisfying this need.

At that time there was concentration on such areas as standardised language extensions for real time computing, operating systems, and man-machine interfaces. News of the workshops reached Europe, and a group of active university and industry based real time computer system developers formed Purdue Workshops Europe. The intention was that Europeans could attend their own local series of workshops, covering topics similar to the USA meetings, and from time to time joint meetings would be held to exchange views. Later, a Japanese Purdue Workshop was set up.

The work of these three continued in parallel for some years. Then the Commission of the European Communities took an interest in the European Group, providing meeting facilities and paying travel and subsistence for Brussels meetings. The name of the European Group was then changed to the European Workshop on Industrial Computer Systems (EWICS). However contact with Purdue Workshops was maintained.

EWICS had until 1983 a number of technical committees which dealt with topics such as Real Time FORTRAN, this group had links with the Instrument Society of America and the main FORTRAN language standardisation work in the American National Standards Organisation (ANSI). Other language work included Real Time BASIC, and the Long Term Procedural Language (LTPL). The group working on LTPL had an early influence on the work towards the ADA language principally through the French membership. Further technical committees dealt with Man-Machine Interface, Operating Systems, Specification, Verification and Validation. However in 1983 the European Community funding largely came to an end, and many of the technical committees ceased to meet.

THE COMMITTEE ON SAFETY SECURITY AND RELIABILITY

The seventh EWICS technical committee to be formed (TC7), was set up to deal with generic issues in Safety, Security and Reliability of industrial real time computers. Beginning work in the mid 1970's on the initiative of Professor R. Lauber at the University of Stuttgart, this group of European experts met four times per year to develop pre-standardisation documents. One of these meetings was the annual meeting for all EWICS and offered an exchange of technical information between the various committees. The other meetings concentrated on developing guidance documents in the form of EWICS TC7 position papers.

The current working pattern of EWICS TC7 owes much to its first Chairman, Professor Lauber. Over the years to early 1985, a number of sub-groups worked towards position papers, and these were constructed from over 400 internal working papers. Some sub-groups decided that their work was so near to the state of the art that it was not yet ready for standardisation and so ceased. others sub-groups presented their findings at Workshops and Conferences. The first EWICS TC7 co-sponsored event was SAFECOMP '79 held at Stuttgart University. Other sub-groups produced position papers and these have been made available on request to European industry, and more recently some have been published. Six position papers were completed by the start of 1985 (1-5).

Working in the pre-standardisation field, EWICS TC7 sought to establish close links with the various standardisation bodies in Europe. The other EWICS technical committees working on languages had already made strong contact with the International Standards Organisation Technical Committee 97 (ISO TC97). However its remit did not sufficiently cover that of EWICS TC7, and other outlets were sought. Through members of EWICS TC7, contact was established with some of the working groups of the International Electrotechnical Committee, particularly with TC45 and work in the nuclear instrumentation and control field. The EWICS TC7 guidelines on Software were largely adopted in working group WGA3 of TC45, and appeared in modified form in 1984 as an IEC standard (6). This success led to further opportunities in IEC groups, such that IEC has now become the main external standardisation body user of EWICS TC7 documents.

When the funding from Europe ceased at the end of the 1979-83 multi-annual programme, EWICS TC7 members decided in consultation with their sponsoring companies that the work was important and should continue. It appeared that Europe might in the near future be in a position to re-commence partial funding, so EWICS TC7 continued to meet and work funded entirely by its members and their companies.

In response to lengthy discussions and a proposal from EWICS TC7, in 1985 an offer was received from the Community to fund a contract to continue an EWICS TC7 work programme. However, this led to a problem for EWICS TC7, since it was not constituted to accept such a contract. Therefore a partner was sought to take the contract, and to manage the work programme. The partner needed to be independent, known to the Community, and have technical relevance to the EWICS TC7 work. The Safety and Reliability Society emerged as the only organisation meeting these criteria.

THE SAFETY AND RELIABILITY SOCIETY

The Safety and Reliability Society was formed in 1980 by a group of people active in the quantitative assessment of safety and reliability (7). They were concerned that no single institution dealt primarily with reliability engineering in its widest sense and few dealt with one of the key uses of this discipline in safety or hazard analysis.

The Society has over the years grown in size and stature. It runs a series of training courses which are the prototype of an accreditation system. Each year a Symposium is held on a topical theme and is well attended. Active Branches have been formed in the UK in London, the North West, the South West and the Midlands and these run extensive winter meeting programmes. It has members who represent virtually all engineering, academic and government authorities in the fields of safety and reliability in the UK and many from overseas. The Society publishes a quarterly Journal, Annual Yearbook (7) and the Proceedings of its Symposia. It collaborates with other professional societies in running events within its sphere of interest.

The main aims of the Society are:-

- To provide a central organisation dedicated to the stimulation and advancement of safety and reliability technology

- To provide an international forum for the exchange of information on safety and reliability engineering

- To establish professional and educational standards for safety and reliability engineers

- To establish standards techniques and encourage consistency in their application

- To encourage organisations and government departments to apply safety and reliability engineering techniques

- To enhance the status of Society members.

The Society welcomed the opportunity to work with EWICS TC7 and the Commission of the European Communities in the furtherance of harmonisation and standardisation in the field of safety and reliability of industrial computer systems. It saw this as a natural and timely development and fully in accordance with its aims.

The Society, through its related company, has accepted the contract from the Commission. In meeting the deliveries require in the contract, the Society has agreed with EWICS TC7 that the established method of working will in the first instance be used to produce documents to the EWICS TC7 proposed programme. The contract provides for a greater freedom in selecting the location of EWICS TC7 meetings, and in supporting more aspects of the work programme.

CURRENT WORK OF EWICS TC7

The current work programme commenced in 1985 and will complete in 1987 with the delivery of four guidelines to the Commission of the European Communities (18), (19), (20), (21). At the time of writing this paper, the first drafts of the guidelines are being prepared, therefore the following descriptive sections represent current views which may change as the drafting progresses.

EWICS TC7 work is now generally system oriented, this can be contrasted with the earlier work which aimed to deal separately with the hardware and software aspects of industrial computer systems.

In addition to these guidelines being system oriented, EWICS TC7 has a policy to issue generic guidelines applicable to a wide range of computer applications in many industries. In this respect the work is compatible with the aims of the Collaborative Project on PES Assessment (8), and particularly Step 7 in the Guidelines Framework (9).

System Integrity

This guideline will address the operational phase of the computer system life cycle (18). It will thus concentrate on how to keep the system safe and reliable whilst it is subject to varying stresses due to its working environment, operational procedures and practise, maintenance and enhancement, and changing requirements. A safety related computer system will require careful monitoring until it is retired from service. The guideline also will define the data to demonstrate that performance is as required, and to support the maintenance and upgrading policies appropriate to safety related computers.

A collection and assessment of methods of keeping systems secure whilst continuing to operate with an acceptable level of safety and reliability will draw on the experience of the EWICS TC7 membership. A number of confidential case studies have already been performed using the operating experience of computer systems used in the transport, chemical and energy industries.

Software Quality Assurance and Metrics

This guideline (19) has two main objectives. It will seek to identify software metrics which best correlate with the software qualities which are sought for in achieving a safe and reliable system. Also the possible usage of software metrics in software quality assurance for the development of safe and reliable systems will be assessed.

This work is particularly difficult, is very near the frontiers of research, and some workers in this field would refute the utility of existing software metrics, discard all, and start anew. However EWICS TC7 seeks to provide good practical guidance rather than to advance the state of the art in the field of software engineering. Through the national and industry balanced membership, drawing on industrial use experience, current research by some members, and the interface with the standards bodies for example the IEEE (10), EWICS TC7 seeks to set down in a useful manner the current best practices.

Currently software metrics are being identified and classified for quality assurance. This is based on case studies and results obtained from the application of software quality standards.

Design for System Safety

Continuing the orientation towards the complete system, this part of the EWICS TC7 programme is reviewing and evaluating design techniques which can be applied to both hardware and software throughout the system life cycle to ensure that safety criteria continue to be met in the presence of faults (20).

There are strong links here with the body of work on Fault Tolerant Computing, and EWICS TC7 is fortunate to have a number of prominent workers in this field who attend and correspond. A number of joint events have been held with the International Federation for Information Processing (IFIP) Working Group 10.4, including SAFECOMP '83 (11) and SAFECOMP '85 (12), and the planned SAFECOMP '86 at Sarlat, France, in October 1986.

All the SAFECOMP events in 79 (Federal Republic of Germany), 81 (USA), 83 (UK), 85 (Italy), 86 (France) have been sponsored by various committees of the International Federation for Automatic Control (FAC), the National IFAC Member Body of the country hosting the event, and always EWICS TC7 have provided the core of the International Programme Committee and many of the technical presentations. Various National Professional

Societies also participate, for example SAFECOMP '83 was co-sponsored by the Institution of Electrical Engineers, the British Computer Society, and the National Centre of Systems Reliability.

Reliability and Safety Assessment

Many of the existing and planned guidelines of EWICS TC7 are directed to the effective means of achieving, or having first obtained the required safety and reliability, retaining the characteristics throughout system life. The new work on reliability and safety assessment has three themes.

First is a study of how computer-based systems can be shown to meet their required level of safety and reliability. In other words, taking account of all the guidance available, the various standards, good engineering practice, what performance will a computer system achieve? How can this be measured in such a way that it is useful, and can it be predicted? Again this work is approaching the boundaries of knowledge particularly for software.

The second theme is to review and evaluate how to specify the safety and reliability criteria for a system, and how to translate them into criteria for system design. The various criteria in use for PES were researched in the Collaborative Project (13), and this together with members input from the industries represented in EWICS is providing a solid base for examining existing safety criteria.

The third theme is to review the techniques for reliability and safety assessment of the system development process and to elaborate a questionnaire for qualitative assessment of the system. Inputs here include the HSE Draft Guidelines on PES (14-16) and in their final format when published (17).

METHOD OF WORKING AND MEMBERSHIP OF EWICS TC7

The working method of EWICS TC7 has evolved over the years. It reflects the interests and needs of its members, the various funding bodies, the many interactions with outside organisations, the programme of work and the production of position papers.

There are usually four main meetings per year at various locations throughout Europe. Individuals and small editorial groups work in-between main meetings on the documents and on gathering and reviewing information. Members and their companies arrange, provide and host facilities for most meetings.

Members time to attend EWICS TC7 meetings and to work between meetings is donated by their employers, the technical programme is therefore designed to assist these employers to further the use of computers in industry in safe and reliable applications. The programme is revised from time to time in committee. For the current programme of work, about 89% of the total cost is donated by industry, and 11% is covered by the contract operated by the Safety and Reliability Society with funds from the European Communities Information Technologies and Telecommunications Task Force.

Whilst the majority of the effort in meetings goes towards the deliverable documents required under the contract, time is set aside for technical presentation and discussion of member projects. This forms a useful informal exchange of research material, sound practical experience and updating of technical information and is highly valued by the members. All members are encouraged to take part, and this provides an excellent vehicle for the members of many nationalities present to get to know each other, and often to be able to informally consult another member who has valuable experience to offer. Recent topics presented by members have included the use of Expert Systems to support safety and reliability assessment, the use of PROLOG in system validation, and experience with the licensing of safety related computer systems.

The main structure of the meetings is to have plenary sessions which include routine business, approval of position papers, and the technical presentations. The majority of the time is allocated to sub-group meetings which work intensively on chosen technical topics. Currently there are four main sub-groups which deal each with one of the current work areas outlined earlier, and each has smaller sub-sub groups of four or five people which work efficiently at a very detailed level.

Each sub-group has an elected chairman, and a volunteer secretary. TC7 as a whole has a secretary, an elected Chairman and Vice-Chairman, and I am currently the Vice-Chairman. All these are the Officers of EWICS TC7, and meet as a management group to monitor technical direction and progress towards the deliverables.

EWICS TC7 membership is largely drawn from users of computers, and these can vary from small to very large companies, including some of the largest European nationalised organisations and public companies. Anyone interested in EWICS TC7 and active in the fields covered by the current contract could be eligible to take part in the work. There is no restriction on industries, companies or national boundaries. Current membership comes mainly from the larger European Community countries (68%), but there are 10% from other Community countries, 17% from other non-Community countries including Eastern Europe, and 5% mainly corresponding members from outside Europe.

Existing members represent a wide range of interests including:-

- Chemical Industry
- Consultants, Computer Systems
- Consultants, Reliability
- Consultants, Safety
- Consultants, Software
- Control Systems Manufacturers
- Defence Industries
- Educational Institutes
- Electricity Generation, non-nuclear
- Electricity Transmission Systems
- Government Departments
- Instrumentation Systems Manufacturers
- Nuclear Chemical Industry
- Nuclear Power Generation
- Railways
- Research
- Road Transport

- Safety Licensing Authorities
- Software Houses
- Telecommunications
- Training

Visitors may attend EWICS TC7 meetings, at the invitation of the Chairman, and at their own expense. A limited number of new memberships with paid travel and subsistence can be offered to experts who accept and complete work items, have experience in the appropriate fields, represent a - country/user industry/standards liaison - combination not currently covered by EWICS TC7 members

INDUSTRIAL USE OF EWICS GUIDELINES

Since the work of EWICS depends on the active involvement of its members, they have the unenviable task of justifying their participation to their own management. To place this in context, one member was chosen as an example. Resulting from discussion with the member, the following section was prepared. The views expressed are my interpretation of that discussion.

The member's company had thought carefully before adopting their current strategies towards standards and in achieving commonality. They saw two basic methods of developing standards, first as exemplified by the General Motors approach to the MAP initiative, where users dictate requirements and on this are based the standards to which suppliers and others conform. This method is supposed to lead quickly to standardisation and then products which conform to the standards. However the member's company chose the consultative method, in which all parties can explore, discuss and achieve results. Despite this being a slower method, it has the advantage of being well tried and allows a conservative but continuous development of standards and equipment.

Therefore this company sought the means to become actively involved in a body which operated in a consultative mode. In selecting the Body, criteria were established which included:-

- The need for a well defined work programme
- A balanced membership
- A continuity of effort and funding
- Documented evidence of work leading to a common acceptance of safety standards.

In choosing EWICS TC7 as the body meeting the criteria, the support and funding of the European Community was decisive. The company saw the completed work as relevant to its engineering programme, and the current work had potential to fulfil as yet unsatisfied needs.

Of course the company has some criticisms of EWICS TC7. The company is the only company in EWICS TC7 representing their sector of industry. They feel, and are working to improve the situation, that links with the national representatives on European Commission programme and funding committees are not as strong as they should be. The output of EWICS is available to its members, to attendees of SAFECOMP, and to the National and International Standards Bodies, but the company sees this material as directly relevant to Industry and feels that wider more immediate publication is necessary.

This concern has also been expressed by other members, and the Safety and Reliability Society together with EWICS TC7 is seeking solutions.

LINKS WITH OTHER WORK

The documentary output of EWICS TC7 is in the form of pre-standardisation guidelines that are of immediate use to industry, and can be used as the basis of full International and National standards. EWICS has long established links with a number of International Standards Organisation (ISO) Technical Committees, but TC7 has developed its strongest international standardisation links with the International Electrotechnical Committee (IEC) having found it the most directly relevant to industrial computer safety and reliability.

The prime link with the IEC and ISO is via members of their technical committees and working groups who are also members of TC7. In this way the following topics are covered:-

IEC TC44 Electrical Equipment of Industrial Machines

- WG1 Aims and means of standardisation of electrical equipment of industrial robots and processor control systems

IEC TC45 Nuclear Instrumentation
SC45A Reactor Instrumentation

- WGA3 Programmed digital computers important to safety in nuclear plants

IEC TC47 Semiconductor devices and integrated circuits
SC47B Microprocessor systems

IEC TC58 Reliability and Maintainability

- WG10 Software Aspects

IEC TC57 Telecontrol, teleprotection and associated telecommunications for electric power systems

IEC TC65 Industrial Process Measurement and Control
SC65A Systems Considerations

- WG6 Standard performance specification for programmable controllers

- WG8 Evaluation of the integrity of system functions

- WG9 Study on Safe Software

SC65B Digital data communications for measurement and control systems

IEC TC74 Safety of data processing equipment and office machines

IEC TC83 Information Technology Equipment

- Functional Safety of IT Equipment

EWICS TC7 also attends some working groups of ISO TC97 and ISO TC184.

At the National level, there is a further set of links via common memberships of various committees and working groups of the National Standardisation Bodies such as BSI, DIN, AFNOR. This may be via membership on the national shadow committees to the relevant IEC and ISO committees, but also provides a wider coverage of the work on safety and reliability standards that apply to industrial use of computers.

At a different National level there are many close working relationships between EWICS TC7 members and the organisations responsible for industrial safety such as the UK Health and Safety Executive and the Federal Republic of Germany TUV's.

The links with IFAC, IFIP, professional societies, and bodies such as the Institution of Electrical and Electronic Engineers which have strong standardisation activities will be maintained.

The links with the Safety and Reliability Society also increase as common membership grows, and the working relationship on the contract develops. The Society approves a long term relationship with the Commission and EWICS TC7 as part of its drive towards professional standards in safety and reliability.

In Europe, EWICS TC& will be seeking links with CEN/CENELEC. It is intended that the new guidelines will be structured and have a content which permits them to include a conformance clause.

A new sub-group is being formed in TC7 to cover the increasing number and complexity of the links that have been described. This sub-group will have the task to service these links, and make EWICS TC7 work more readily accessible.

CONCLUSIONS

EWICS TC7 was set up to propose schemes, principles, procedures and guidelines to international bodies. In this it has succeeded to the extent that it now serves as a unique forum in Europe for the harmonisation between industries, countries and standards bodies.

The means of achieving this harmony is the exchange of experiences and ideas concerning safety that takes place between the expert members, and the application of the results of the EWICS work of EWICS TC7 in their own industries.

The adoption of the work of EWICS TC7 is now widespread in standards bodies. These links are placing an increasing load on individual members, who are in demand to attend working group meetings.

The development of professional standards in the industrial computer system safety and reliability field is enhanced by the links between EWICS TC7 and the Safety and Reliability Society. The work of EWICS TC7 is becoming more readily available to the reliability and safety analysts.

ACKNOWLEDGEMENT

This paper could not have been written without the support of a large number of other people. I must thank all the members of EWICS TC7, but particularly Stuart Nunns who in discussion allowed me to reflect on the views of a large UK industrial user of computer systems and the role of standards. Also I must thank the Safety and Reliability Society for the extracts from the 1986 Yearbook.

The views expressed in this paper are those of the author.

REFERENCES

1. Guideline for Verification and Validation of Safety Related Software. European Workshop on Industrial Computer Systems. Computers and Standards, Vol. 4, No. 1, 1985. North Holland. (Position Paper No. 3).

2. Safety Related Computers: Software development and systems documentation. European Workshop on Industrial Computer Systems. Published by Verlag TUV Rheinland GmbH, Cologne, 1985. (Position Papers Nos. 1 and 4).

3. Hardware of Safe Computer systems, Position Paper No. 2, European Workshop on Industrial Computer Systems, June 1982.

4. Techniques for Verification and Validation of Safety Related Software, Position Paper No. 5, European Workshop on Industrial Computer Systems, January 1985.

5. System Requirements Specification for Safety Related Systems, Position Paper No. 6, European Workshop on Industrial Computer Systems, January 1985.

6. IEC Standard Software for safety related Nuclear Instruments.

7. The Safety and Reliability Society, Yearbook 1986, PO Box 25, Cambridge Arcade, Southport, UK.

8. Bell, R., Assessment architecture and performance of Industrial Programmable Electronic Systems (PES) with particular reference to Robotics. See this Symposium.

9. Daniels, B. K., Guideline Framework for the Assessment of PES. See this Symposium.

10. Draft Standard, Measures for Reliable Software, IEEE P982, March 15 1985.

11. SAFECOMP '83, Proceedings of the Workshop, Cambridge, UK, September 1983. ed. Baylis, J. A., Pergamon Press.

12. SAFECOMP '85, Proceedings of the Workshop, Como, Italy, October 1985. ed. Quirk, W. J., Pergamon Press.

13. CEC Collaborative Project, Assessment, architecture, and Performance of Industrial Programmable Electronic Systems with particular reference to robotics safety, Final Report, 1986.

14. Guidance on the safe use of Programmable Electronic Systems: Part 1: General Requirements, draft document for consultation, Health and Safety Executive, UK, July 1984.

15. Guidance on the safe use of Programmable Electronic Systems: Part 2: Safety Integrity Assessment, draft document for consultation, Health and Safety Executive, UK, July 1984.

16. Guidance on the safe use of Programmable Electronic Systems: Part 3: Safety Integrity Assessment Case Study, draft document for consultation, Health and Safety Executive, UK, July 1984.

17. Bell, R., Guidance on the use of PES in safety related applications. See this Symposium.

EWICS GUIDELINES IN PREPARATION

18. Systems Integrity, Position Paper, European Workshop on Industrial Computer Systems, due end 1987.

19. Software Quality Assurance and Metrics, Position Paper, European Workshop on Industrial Computer Systems, due end 1987.

20. Design for System Safety, Position Paper, European Workshop on Industrial Computer Systems, due end 1987.

21. Reliability and Safety Assessment, Position paper, European Workshop on Industrial Computer Systems, due end 1987.

GUIDANCE ON THE USE OF PROGRAMMABLE ELECTRONIC SYSTEMS IN SAFETY RELATED APPLICATIONS

R BELL

Health and Safety Executive
Bootle
Merseyside, L20 3QZ UK

INTRODUCTION

In August 1984 the Health and Safety Executive (HSE) carried out an extensive consultation exercise aimed at gaining agreement on 3 documents concerned with the safe use of programmable electronic systems (PESs). It is hoped to publish these guidance documents, titled **"Guidance on the use of programmable electronic systems in safety related applications: Parts 1-3"**.

The proposed guidance documents provide generically based guidance that will enable the safety integrity of systems incorporating PESs to be determined irrespective of the application. However, it is recognised that there is a vast spectrum of PES applications of wide ranging complexity and hazard potential and it is envisaged that further, second-tier documents, will be required in specific areas (eg machinery safeguarding, safety shutdown systems). The framework will enable a consistent approach to be adopted in the second-tier guidance and result in simplified but more specific advice. This paper outlines the philosophy and the essential safety principles contained in these proposed generically-based guidance documents.

NOTE: The term "safety integrity" refers to the ability of a safety related system to perform its safety functions correctly when required. In determining the safety integrity, all causes of failure which lead to an unsafe state should be included.

SYSTEMS UNDER CONSIDERATION

The documents are concerned with those PESs which either acting alone or in combination with non-programmable systems, provide the requisite level of safety. Such systems are referred to as safety related systems and are those systems upon which the safety of the plant/machine depends and whose failure would be included in the events leading to the hazard or hazards in question.

The guidance in the proposed documents does not apply if an adequate level of safety is assured by separate non-programmable systems. The safety integrity of these non-programmable systems should not be inferior to those which have been accepted as good engineering practice in situations with a potential for hazard of similar severity.

The PES refers to systems incorporating a computer and extends from plant/machine sensors, or other input devices to the plant/machine actuators or other output devices. Diagrams illustrating PES structure and notation are given in Figure 1.

PROBLEM AREAS

In formulating guidance on PESs cognizance has to be taken of the following:

1. It is rarely possible that the software can be fully tested under all possible operating conditions and so program faults may remain unrevealed until a particular set of operating condition causes a program failure.

2. A fault may be introduced into a program or data stored in the memory in the PES as a result of some transient fault or disturbance.

3. The failure modes of a PES are significantly more complex than for conventional control systems and are not always predictable.

4. Particular attention needs to be paid to the electromagnetic compatibility (EMC).

5. The 'experience phase' with any specific system is limited because of the relative ease of re-programming and the rapid development of hardware technology. This does not easily allow data to be accumulated on a system basis for reliability/safety integrity assessment purposes.

6. Particular attention, at both design and operational stages, has to be paid to minimise both inadvertant changes, and deliberate unauthorised changes, to the software. These problems are exacerbated by the difficulty of software assessment after a change has been made.

DESIGN STRATEGY

The design strategy for safety related PESs is based upon the safety related systems meeting certain **safety principles** with respect to three main elements:-

a. The **configuration.**

b. The **reliability** of the hardware with respect to random, independent failures in dangerous mode of failure.

c. The safety integrity with respect to non-quantifiable, qualitative aspects. That is, the **quality** of the procedures used in specification, design, implementation and operation.

These three elements are hereafter referred to by the terms **configuration, reliability and quality.** The exact package of safety measures comprising a. b. and c. above will depend upon the level of safety to be achieved and the particular application.

Configuration requirements

This element in the achievement of the required level of safety integrity, primarily relates to reliability with respect to both random hardware failures and systematic failures, particularly those due to software errors.

There are three principles underlying the guidance on what constitutes an acceptable configuration:

a. The combined number of PES and non-PES based safety related systems which are capable, independently, of maintaining the plant in a safe state when required should not be less than the number of conventional systems which have traditionally been accepted as good engineering practice in situations with the potential for a hazard of a similar severity.

b. No single failure of hardware in any of the programmanble electronics should cause a dangerous mode of failure of the total configuration of safety related systems. It should be recognised that systematic hardware failure causes may affect all identical designs of programmable electronics.

c. No single failure of software in any of the programmable electronics should cause a dangerous mode of failure of the total configuration of safety related systems. It is assumed that the failure cause will affect all identical software.

Reliability requirements

This element in the achievement of the required level of safety integrity relates to reliability with respect to random hardware failures in a dangerous mode of failure. The underlying principle is that, taking the combined effect of each safety related system, the overall failure rate in the dangerous direction, or, for a protection system the probability of failure to operate on demand, should not be inferior to the equivalent conventional systems which have been accepted in situations with a potential for a hazard of a similar severity.

There are 3 principal means of determining whether this criterion has been satisified (see a. b. and c. below).

a. A qualitative assessment using engineering judgement of both the sensors and actuators on the plant/machine and of the programmable electronics.

b. A quantified assessment of the reliability of the safety related systems.

c. An overall quantified assessment of the plant/machine to ensure that the frequency of the hazard will not be greater to that accepted in the past for hazards of comparable severity.

Quality requirements

This element in the achievement of the required level of safety integrity relates to the precautions taken against:

- errors or omissions in the requirements specification of the safety related system(s).

- systematic failures from all causes including software

Quality aspects must be considered at every stage in the specification, design, construction, testing, commissioning, operation, maintenance and modification of the PES, in relation to both hardware and software. The level of quality is determined by the procedural and engineered measures for the avoidance of errors in design and other defects which could cause systematic failure.

The level of quality required depends primarily upon the consequences of failure and important pointers can be taken from the procedural and engineering measures traditionally considered approproriate for conventional systems associated with similar hazards.

The minimum level of quality, which will be sufficient for those situations in which the most severe consequence of failure will be a minor injury, should satisfy the criteria set out below:-

a. Quality of manufacture. The programmable electronics should be manufactured to an established quality assurance system. Other parts of the PES should be to a level of quality not inferior to that which has been achieved, in situations with the potential for hazards of a similar severity, by conventional safety related systems which have traditionally been accepted as good engineering practice.

b. Quality of implementation. The safety related systems particularly those that are PES based, should be engineered by competent and experienced persons having an understanding of the safety engineering principles of PESs (such as the guidance contained in the proposed documents).

For more serious hazards, in addition to the above, each procedural and engineering aspect of the PES should be carefully and systematically examined to ensure an appropriate level of quality. Part 2 of the proposed guidance documents provides a method for carrying out a qualitative examination by means of checklists covering various aspects of the system which should be considered at each stage from specification through to operation and maintenance.

DESIGN ASSESSMENT FRAMEWORK

Before PES-based safety related systems can be used it is necessary for those who select, apply, program and use PESs to ensure that the problem areas have been overcome and the required level of safety integrity is met for the application in question. The design should incorporate the **safety principles** outlined previously. A framework within which these safety considerations can be examined systematically is outlined below.

a. Analysis of the Hazards

i. identify the potential hazards
ii. evaluate the events leading to these hazards

b. Identify the safety related systems. That is, those systems whose failure are included in the events leading to the hazards identified in a. above.

c. Decide on an acceptable level of safety integrity for the safety related systems.

d. Safety integrity analysis

i. assess the level of safety integrity achieved by the safety related systems

ii. compare d.i. and c. to ensure that the level of safety integrity achieved is at least as good as that required.

The required level of safety integrity (item c. above) of systems having a safety role will depend upon many factors, eg severity of injury, number of people at risk, frequency at which a person or persons are exposed to a risk and the duration of exposure.

In designing any system having a safety role, it is a pre-requisite that the required safety integrity level be known. This has traditionally, in some cases, for non-programmable systems, been specified in quantified terms, eg the probability of failure on demand, the dangerous failure rate of a specified safety function or the acceptable frequency of occurrence of a hazard. In many cases, however, the design safety integrity will be based upon a set of wide ranging qualitative criteria. In the context of safety related systems, one or more of which is PES based, the required safety integrity will be based upon a package comprising the three elements of the **safety principles**.

The objective should be to ensure that the safety integrity of the total configuration of PES and non-PES safety related systems should not be inferior to that which has been achieved, in situations with the potential for hazard of similar severity, by conventional safety related systems which have traditionally been accepted as good engineering practice.

APPLICATION OF SAFETY PRINCIPLES

Table 1 illustrates the application of the **safety principles** in the context of protection systems, in order to meet the above objective (see previous paragraph).

SUMMARY

The Health and Safety Executive hope to publish in the near future 3 documents titled **"Guidance on the use of programmable electronic systems used in safety related applications Parts 1-3"**. The proposed documents provide a framework within which the safety integrity of PES based safety related systems can be determined.

It is recognised that there is a vast spectrum of PES applications of wide ranging complexity and hazard potential. The documents will enable further guidance to be formulated in specific areas (eg machinery safeguarding, safety shut-down systems). The development of this **second-tier guidance,** within the framework of these grenerically based documents, will enable the application in question to be specifically addressed and should result in simplification of the guidance and lead to a consistent approach being adopted in those areas where guidance is required.

Figure 1: Diagram to Illustrate the PES Structure and Notation

Figure 1a: Basic PES Structure

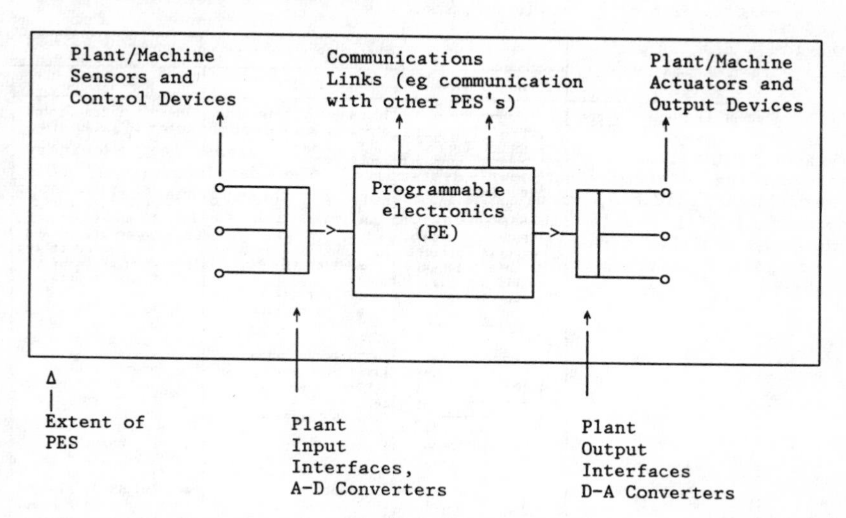

Figure 1b: Diagrammatic representation of a PES with single Programmable Electronics

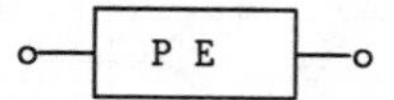

PE = Programmable Electronics

Figure 1c: Diagrammatic representation of a PES with dual Programmable Electronics but shared sensors and actuators

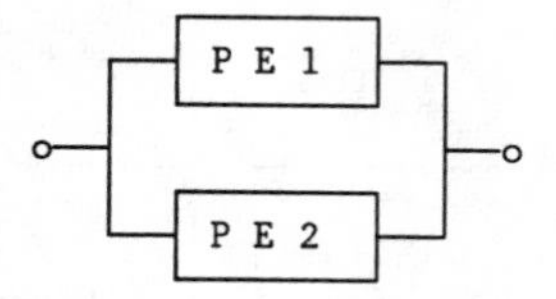

Figure 1d: Diagrammatic representation of a system based on conventional, non programmable hardware

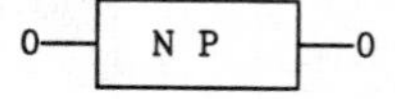

NP = conventional non-programmable hardware

TABLE 1: EXAMPLES OF THE APPLICATION OF THE SAFETY PRINCIPLES: PROTECTION SYSTEMS

NOTE: PE = programmable electronics

CONVENTIONAL CONFIGURATION	EQUIVALENT PES BASED SYSTEM(S)		
	CONFIGURATION	RELIABILITY	QUALITY
CASE 1: Single System PROTECTION	PE1 PROTECTION / PE2 PROTECTION (in parallel) PE1 and PE2 based on diverse software and, unless certain conditions apply, diverse hardware or PE PROTECTION / NP PROTECTION (in parallel)	There are 3 principal means of ascertaining that the reliability requirements are satisfied:- a) qualitative appraisal b) quantified assessment of the PES c) overall plant safety assessment In this case, the level of hazard is likely to be low and (a) is likely to be most appropriate.	There are 2 principal means of satisfying the quality requirements:- a) The system should be manufactured to an established QA/programme and engineered taking cognizance of these documents or b) As (a) above. In addition, the system should be systematically examined using the specified, or equivalent, method. In this case, the level of hazard is is likely to be low and (a) is likely to be most appropriate.
CASE 2: Two Systems PROTECTION PROTECTION See comments for Case 1.	PE1 PROTECTION PE2 PROTECTION or PE PROTECTION NP PROTECTION	The level of hazard is likely to be more serious than in Case 1 and, although (a) above may be sufficient, (b) or (c) may also be appropriate.	The level of hazard is likely to be more serious than in Case 1 and (b) above may be appropriate.
CASE 3: Three systems PROTECTION PROTECTION PROTECTION	* PE1 PROTECTION PE2 PROTECTION PE3 PROTECTION See Comments for Case 1 above.	The level of hazard is likely to be high and (b) or (c) above are likely to be appropriate.	The level of hazard is likely to be high and (b) above will be necessary. A high degree of compliance with the examination criteria will probably be appropriate.

* NOTE: PE3 may be identical to PE1 or PE2; overall system option incorporating non-programmable elements not shown.